What Can I Do Now?

Preparing for a Career in the Environment

Ferguson Publishing Company, Chicago, Illinois

Printed in the United States of America
V-4

Library of Congress Cataloging-in-Publication Data

Preparing for a career in the environment.
p. cm. -- (What can I do now?)
Includes bibliographical references and index.
Summary: Introduces the environmental industry, presents eight promising careers in theat field and ways to prepare for them, and discusses immediate ways to get involved, including internships and volunteerism.
ISBN 0-89434-249-5
1. Environmental sciences--Vocational guidance--Juvenile literature. [1. Environmental sciences--Vocational guidance. 2. Vocational guidance.]
II. Series.
GE60.P74 1998
363.7'0023--dc21 98-16920
CIP
AC

Ferguson Publishing Company
200 West Madison, Suite 300
Chicago, Illinois 60606
800-306-9941
www.fergpubco.com

About the Staff

- **Holli Cosgrove,** *Editorial Director*
- **Andrew Morkes,** *Editor*
- **Veronica Melnyk,** *Assistant Editor*
- **Patty Cronin, Veronica Melnyk, Beth Oakes, Kate Quinlan, Elizabeth Taggart,** *Writers*
- **Connie Rockman, MLS; Alan Wieder,** *Bibliographers*
- **Patricia Murray, Bonnie Needham,** *Proofreaders*
- **Joe Grossmann,** *Interior Design*
- **Parameter Design,** *Cover Design*

Contents

Introduction

There are many people just like you who want to get involved with making the Earth a cleaner, healthier, and more pleasant place to live. You may see an environmental career in your future and wonder how you can get started right away—while still in high school. There are countless areas of the environment you can work in—areas that you can match your skills and talents to. All you need is a general interest in the field to begin with. Some of you may feel that you need a college degree before you can get started in this field, but many environmental jobs require only a two-year degree after high school—and you can enter some right after you graduate. There are also many careers for the advanced, motivated student. Jobs are out there for the high school-to-work crowd and for those of you who want to pursue advanced degrees. So don't let excuses get in the way.

There is absolutely no reason to wait until you get out of high school to "get serious" about a career. That doesn't mean you have to make a firm, undying commitment right now. Gasp! Indeed, one of the biggest fears most people face at some point (sometimes more than once) is choosing the right career. Frankly, many people don't "choose" at all. They take a job because they need one, and all of a sudden ten years have gone by and they wonder why they're stuck doing something they hate. Don't be one of those people! You have the opportunity right now—while you're still in high school and still relatively unencumbered with major adult responsibilities—to explore, to experience, to try out a work path. Or several paths if you're one of those overachieving types. Wouldn't you really rather find out sooner than later that you're not cut out to be an oceanographer after all, that you'd actually prefer to be an environmental lobbyist? Or a national park service employee?

There are many ways to explore the field of the environment. What we've tried to do in this book is give you an idea of some of your options. The chapter "What Do I Need to Know about the Environment?" will give you an overview of the field—a little history, where it's at today, and promises of the future; as well as a breakdown of its structure—how it's organized—and a glimpse of some of its many career options.

The "What Do I Need to Know about Careers?" section includes six chapters, each describing in detail a specific environmental career: ecologist, environmental lobbyist, groundwater professional, hazardous waste management technician, national park service employee, and oceanographer. These chapters rely heavily on first-hand accounts from real people on the job. They'll tell you what skills you need, what personal qualities you have to have, what the ups and downs of the jobs are. You'll also find out about educational requirements—including specific high school and college classes—advancement possibilities, related jobs, salary ranges, and the employment outlook.

The real meat of the book is in the section called "What Can I Do Right Now?" This is where you get busy and DO SOMETHING. The chapter "Get Involved" will clue you in on the obvious—volunteering and interning—and the not-so-obvious—summer camps and summer college study, high school clubs, and student organizations.

In keeping with the secondary theme of this book (the primary theme, for those of you who still don't get it, is, "You can do something now"), the chapter "Do It Yourself" urges you to take charge and start your own programs and activities where none exist—school, community, or even national. Why not?

While we think the best way to explore the environment is to jump right in and start doing it, there are plenty of other ways to get into the environmental mind-set. "Surf the Web" offers you a short annotated list of Web sites where you can explore everything from job listings (start getting an idea of what employers are looking for now) to educational and certification requirements to on-the-job accounts from those who keep the environment safe.

"Read a Book" is an annotated bibliography of books (some new, some old) and periodicals. If you're even remotely considering a career in the environment, reading a few books and checking out a few magazines is the easiest thing you can do. Don't stop with our list. Ask your librarian to point you to more materials. Keep reading!

"Ask for Money" is a sampling of environmental scholarships. You need to be familiar with these because you're going to need money for school. You have to actively pursue scholarships; no one is going to come up to you one day and present you with a check because you're such a wonderful student. Applying for scholarships is work. It takes effort. And it must be done right and often as much as a year in advance of when you need the money.

"Look to the Pros" is the final chapter. It lists professional organizations you can turn to for more information about accredited schools, education requirements, career descriptions, salary information, job listings, scholarships, and more. Once you become a student in an environmental field, you'll be able to join many of these. Time after time, professionals say that membership and active participation in a professional organization is one of the best ways to network (make valuable contacts) and gain recognition in your field.

High school can be a lot of fun. There are dances and football games; maybe you're in band or play a sport. Great! Maybe you hate school and are just biding your time until you graduate. Too bad. Whoever you are, take a minute and try to imagine your life five years from now. Ten years from now. Where will you be? What will you be doing? Whether you realize it or not, how you choose to spend your time now—studying, playing, watching TV, working at a fast food restaurant, hanging out, whatever—will have an impact on your future. Take a look at how you're spending your time now and ask yourself, "Where is this getting me?" If you can't come up with an answer, it's probably "nowhere." The choice is yours. No one is going to take you by the hand and lead you in the "right" direction. It's up to you. It's your life. You can do something about it right now!

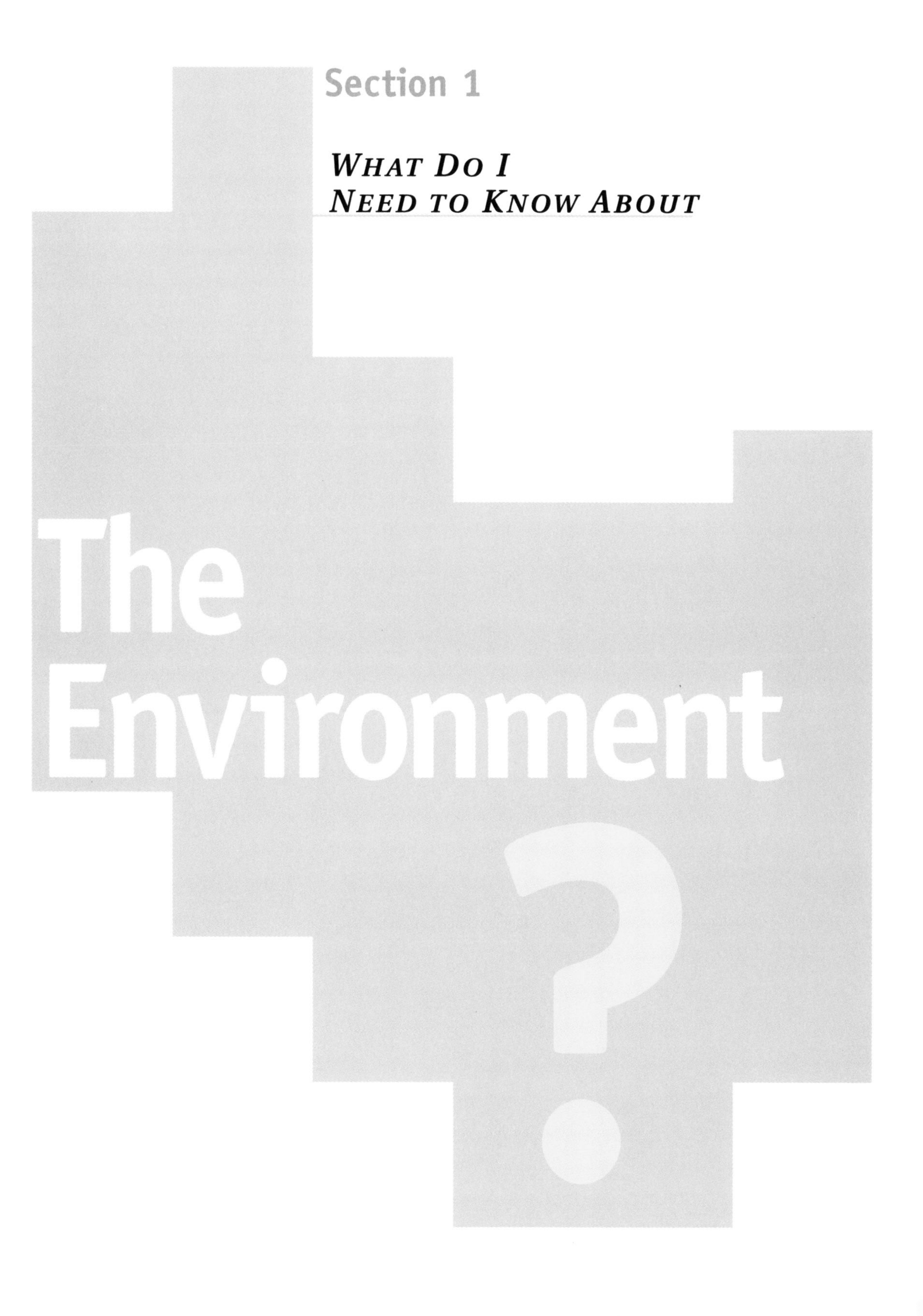

Section 1

What Do I Need to Know About

The Environment ?

What do you think of

when you hear the word environmentalist? A bearded hippie preaching about recycling? A teary-eyed movie star holding a press conference about the destruction of the rain forests? The reality of environmental work today is actually much more diverse and complex than that. Thousands of scientists, engineers, and naturalists—along with health care workers, technicians, business owners, information management specialists, public relations people, teachers, lawyers, and others—are tackling the many needs of the environment in literally thousands of different ways. Some people who care about the environment work in labs and offices, others argue cases in court or lobby members of Congress, others take soil samples or observe wildlife in the field, still others teach or study in the classroom. And we haven't even scratched the surface of the variety of jobs available to someone interested in this field.

General Information

Environmental careers have exploded in the last twenty-five to thirty years, and just about everybody expects that growth to continue well into the next century. Of course, the reasons behind this "success" often are disquieting, if not ominous. Studies showed almost every U.S. metropolitan area was still plagued by dangerous levels of smog in 1990, with Los Angeles leading the way, despite years of trying to cut down on auto and big industry emissions. The end of the Cold War has revealed some disastrous environmental problems in Eastern Europe and the former Soviet Union, the legacy of decades of heavy industry but little or no pollution control.

Wildlife habitats continue to be threatened, and nearly one thousand plants and animals in the United States alone were considered endangered in 1997, according to the U.S. Fish and Wildlife Service. Periodic contamination, or poisoning, of water supplies by pollutants shows that much work is left to be done to ensure safe drinking water for all.

GENERAL INFORMATION, CONTINUED

Fears about such phenomena as global warming, holes in the ozone, acid rain, and other environmental crises have many people wondering what the planet will be like in the coming decades for their grandchildren or even for their grandchildren's children.

Furthermore, public support for environmental work is strong and apparently continues to grow all the time. When asked in a national poll if they would rather see the environment saved even if prices had to go up, 80 percent of respondents said "yes" in 1991, compared to only 51 percent in 1981.

"Something will have gone out of us as a people if we ever let the remaining wilderness be destroyed . . . if we pollute the last clean air and dirty the last clean streams and push our paved roads through the last of the silence . . . "—Wallace Stegner

The Environmental Careers Organization (ECO) breaks down environmental careers into these broad categories: parks and outdoor recreation, air and water quality management, education and communication, hazardous waste management, land and water conservation, fishery and wildlife management, forestry, planning, and solid waste management. What's more, says ECO, new kinds of jobs are being created all the time to meet new demands. There currently are more than thirty major environment-related areas for study at the college level, and some experts say that we've only begun to see the tip of the iceberg in terms of types of environmental jobs.

Many environmental problems are interconnected and broad-ranging. Acid precipitation, often known as acid rain, is caused by pollutants released into the air from cars, trucks, planes, factories, and power plants. When acid rain or acid snow falls, lakes, streams, and oceans are contaminated, causing harm to marine and freshwater life, drinking-water supplies, crops, buildings, and forests. In other words, air pollution can lead to water pollution and to forest destruction. Likewise, forest destruction leads to air and water pollution. These problems further become human problems when they directly affect our health and well-being. This interconnectedness of environmental problems has led to a distinct trend: environmental professionals increasingly are encouraged to be cross-trained, to be able to work on issues involving, for example, both wildlife biology and forestry.

The history of environmental problems is closely linked to two things: the expanding world population and the many advances in technology that began with the Industrial Revolution. When relatively few people inhabited the earth and before they lived in cities, they could afford to pollute their environment with few adverse effects. Trees could be cut for fuel in small enough numbers, and the smoke from widely scattered fires dissipated in the atmosphere. Sewage and garbage could be disposed of without causing disease. Air and water pollution existed, but not in large enough degrees to spark a public outcry or to halt the activities that caused them.

As early as 1300, London, the most heavily populated city in England, was already suffering ill effects from the widespread use of coal as a fuel. The air was thick with smoke and smelled terribly. In 1306, King Edward I banned coal-burning in London, though the reasons are unclear. Some believe the king was successfully lobbied by the powerful wood establishment to ban coal at a time when England's wood was in short supply, thus keeping wood prices high. Others believe that the king showed remarkable wisdom and foresight in recognizing the harmful effects of air pollution. After a time, however, coal use resumed because there simply wasn't enough wood to fuel the city.

The state of the environment started to deteriorate much more rapidly with the advent of the Industrial Revolution in Europe in the mid-1700s. Large numbers of people living in closely concentrated cities began to produce immense quantities of pollutants with no means of efficiently and harmlessly dispersing them. Great clouds of smoke spewed from factories fueled by coal. By 1750, England had become the first country to be largely powered by coal, but people were so excited by the mechanization of industry and the conveniences that industry and city living afforded, that they pressed full speed ahead without giving much consideration to the damage being done. The mastery of coal, combined with the advent of the steam engine, not only made England the leading industrial nation in the 1800s, but it made it the world's largest polluter as well. Smog engulfed the city, and all manner of wastes were dumped into the Thames River. (Later, when these wastes permeated the drinking-water supply, thousands of people became ill with typhoid fever.) In the 1900s the situation was worsened by continuing technological developments, including the automobile. In 1952, the sulfur dioxide smog was so severe in London that four thousand deaths were attributed that year to breathing and respiratory problems caused by pollution.

General Information, continued

The United States relied on wood as its primary fuel until 1850, as the amount of wood was abundant and clearing the forests was a major American industry. Coal became the most popular fuel from about 1885 until about 1950, when oil and gas ended the reign of coal. Although there was no single precipitating event in this country to draw awareness to environmental problems, the biggest contributing factor was probably the net consequence of our reliance on and high consumption of nonrenewable fossil fuels. The "environmental movement" in the United States was spawned in the 1960s. By that time, most lakes and rivers in this country had suffered damage, and Lake Erie was actually declared dead, devoid of any plant or animal life. (It has since been revived to some extent.) Air pollution plagued every major U.S. city. And scientists discovered in the mid-1960s that fir and spruce trees in the eastern part of the country were being damaged by something called acid rain.

From the beginning of the Industrial Revolution, some voices have called out for protection, conservation, and care of the environment. Henry David Thoreau (1817-1862) was celebrating nature and warning about the encroachment of railroads and industry in the mid-nineteenth century; the Audubon Society and the Sierra Club were formed in the 1890s out of concern that wilderness areas were disappearing; and the National Wildlife Refuge System helped stop the slaughter of Pelican Island, Florida, waterfowl in 1903 (the birds' feathers were being used for women's hats). The National Park Service was created in 1916 to conserve the national parks' scenic areas and wildlife. Fishery management began in the 1870s and was a busy profession by the 1950s. The first Clean Air Act was passed in 1955, the Air Quality Act in 1967. Rachel Carson's (1907-1964) book, *Silent Spring*, published in 1962, was an urgent statement about the serious health and environmental threats of insecticides. But there was no environmental industry per se until the 1960s, according to ECO; a comparative few rangers, foresters, public health officials, and advocates basically made up the environmental field.

However, the watershed year for the U.S. environment movement surely was 1970. In 1970, Americans celebrated the first Earth Day, a nationwide event to promote awareness of the planet's fragility. (To find out more about the founder of Earth Day, Gaylord Nelson, see the sidebar on the next page.) Earth Day called on consumers and government leaders to clean up the country's messes, to develop alternative and renewable energy resources, and to preserve and take better care of the pristine areas that remained. It raised public awareness and helped fire up environmental action at the grassroots level—that is, citizens began to take action themselves to help the environment. The federal government established first the Council on Environmental Quality and then the U.S. Environmental Protection Agency (U.S. EPA) to implement environmental legislation. Out of the U.S. EPA were born all the state-level EPA organizations, which also work to implement the laws.

The National Environmental Policy Act, which took effect in 1970 as well, required that any federal agency with plans to build, support, or regulate

Gaylord Nelson, Founder of Earth Day

On April 22, 1970, more than twenty million people participated in demonstrations, attended teach-ins, and listened to speeches about the environment. It was, at the time, the largest demonstration in history. This watershed event, called Earth Day, inspired our country to recognize the importance of environmental issues. The annual event, which continues to change the world today, grew out of one man's frustration with federal government's inattention to environmental issues. That man was Gaylord Nelson, a senator from Wisconsin.

Gaylord Nelson was born on June 14, 1916. He earned his BA at San Jose State College and his law degree from the University of Wisconsin Law School. He served as first lieutenant in the U.S. Army in Okinawa before returning to Madison, Wisconsin, to practice law. From 1959 to 1962, Gaylord Nelson served as governor of Wisconsin. During his tenure in this office, he initiated a program to purchase one million acres of land for recreation, parks, and wildlife areas. This revolutionary program was funded through a penny-a-pack tax on cigarettes.

When Nelson was elected to represent Wisconsin in the U.S. Senate in 1962, he was troubled by that body's apparent indifference to environmental problems. To draw attention to environmental issues, Nelson urged President Kennedy to embark on a Conservation Tour. The President agreed and embarked on the tour in September of 1963. The tour failed to galvanize the public to environmental awareness and Nelson continued to search for new ways to draw attention to environmental problems.

In the late 1960s, Nelson ran across an article about the teach-ins being held on college campuses. He was electrified by the concept. In a speech in Seattle in September 1969, Nelson announced that there would be a national environmental teach-in in early 1970. The nation's response was overwhelming.

The public's enthusiastic response to Earth Day convinced politicians throughout the country that their constituents were passionately interested in environmental issues. Soon after, Congress created the Environmental Protection Agency. Since the inception of Earth Day, Congress has enacted nearly 40 major federal environmental laws. Today, Earth Day is observed by more than 200 million people worldwide, in more than 140 countries.

Gaylord Nelson was also the sponsor of the 1964 Wilderness Act. He introduced legislation to require automobile fuel efficiency standards, ban the use of DDT and Agent Orange, control strip mining, and ban the use of phosphates in detergents. In 1995, he received the Presidential Medal of Freedom for his lifelong efforts on behalf of the environment. Today, Nelson continues to fight for the environment as the head of the Wilderness Society. His achievements demonstrate that, with enough conviction and energy, one person can make a difference. Together we can save the planet.

General Information, continued

a large facility such as a highway, dam, or power plant first must issue an environmental impact statement (EIS) that would assess any damage the project might cause to the environment. The EIS must also take alternative proposals under consideration. A preliminary EIS is reviewed by federal, state, and local authorities and is made available to the public before the federal agency prepares a final EIS and decides whether to proceed with the project. Sometimes an EIS has resulted in projects being challenged legally and delayed; one well-known example is the Tellico Dam on the Little Tennessee River, which might have eradicated a little-known, endangered species of fish called the snail darter. The project was challenged by eleven lawsuits and held up for several years. It was only resumed after the three-inch-long fish were relocated to other tributaries of the river. The status of the snail darter since has been upgraded from endangered to threatened.

The environmental movement picked up steam in the 1970s, and in the 1980s—despite some pressure from big business and President Reagan—environmental groups, such as the Sierra Club, Audubon Society, Friends of the Earth, and Wilderness Society, gained considerable lobbying power on Capitol Hill, and many laws were passed to clean up and protect the environment. These included the Resource Conservation and Recovery Act of 1976, meant to regulate how industry got rid of its wastes. They also included the Superfund Act for cleaning up toxic wastes at abandoned sites and prosecuting the responsible parties, the Clean Air Act, the Wild and Scenic Rivers Act, the Clean Water Act, and several others.

A stance on the environment, pro or con, now is part of many candidates' platform. For example, former president Ronald Reagan's pro-industry/anti-environmental stance on most issues, and in particular his appointment of anti-environmentalist James Watt as Secretary of the Interior (the department in charge of the National Park system), sparked a great deal of outrage in the environmental community and led to record-breaking memberships in environmental groups. Watt eventually relinquished his post under pressure. On the other hand, during the Clinton administration, some people have been disappointed that Vice President Al Gore, an avowed environmental-

Lingo to Learn

Ecosystem: *A community of various species that interact with one another and with the chemical and physical factors making up its nonliving environment.*

Endangered species: *A species having so few individual survivors left that it may become extinct over all or most of its natural range.*

Environmental science: *The study of how we and other species interact with each other and the nonliving environment of matter and energy.*

Hazardous waste: *Any discarded substance, usually chemicals, that can cause harm to humans.*

Recycling: *The collecting and reprocessing of a resource in order to turn it into a new, reusable product.*

Fast Fact

A recent report released by the United Nations Environment Program states that twenty-five thousand people die daily as a result of poor water quality.

ist, has not been more aggressive on environmental issues.

While environmental quality has improved on many fronts, serious problems remain to be solved. Industrial and auto emissions have been cut significantly—for example, levels of airborne lead dropped 79 percent between 1981 and 1990, and carbon monoxide dropped 36 percent during that time—but many cities still don't meet EPA regulations for allowable limits of airborne pollutants.

"It is only in the most recent, and brief, period of their tenure that human beings have developed in sufficient numbers, and acquired enough power, to become one of the most potentially dangerous organisms that the planet has ever hosted. . . We, humanity, have finally done it: disturbed the environment on a global scale."—Thomas E. Lovejoy

Other problems still to be resolved include the worldwide phasing out of ozone-depleting chemicals in order to protect the damaged ozone layer, which shields us from harmful radiation; the reduction of so-called greenhouse gases, including curbing emissions of carbon dioxide, to reverse the trend of global warming; and putting a stop to the burning of tropical forests, which also contributes to global warming and has many other deleterious effects. Some of these issues were addressed at the international level for the first time in the Earth Summit, a gathering of leaders from 178 nations in Rio de Janeiro, Brazil, in June 1992. The majority of the nations represented signed treaties aimed at curbing global warming and at promoting biodiversity, or protecting endangered plants and wildlife. The delegates at the convention also adopted a nonbinding plan called Agenda 21, which calls for industrial nations to provide financial and technological assistance to developing countries while also encouraging them to protect their environments.

In December 1997, representatives from 159 nations met in Kyoto, Japan, to discuss ways to reduce greenhouse gas reductions, the principle

GENERAL INFORMATION, CONTINUED

agents responsible for global warming. Thirty-eight developed nations, including the United States, agreed in principle to reduce their emissions of the six greenhouse gases—carbon dioxide, nitrous oxide, methane, hydrofluorocarbons, perflurocarbons and sulfur hexaflouride—by an average of 5.2 percent by 2012. The United States, the world's largest producer of greenhouse gases, agreed to reduce emissions to 7 percent below what they were in 1990 (although Congress must ratify the treaty before it can be implemented). While, many consider any agreement a success, others questioned how these voluntary reductions could be monitored and enforced. Many developed countries, such as the United States, were disappointed that developing countries such as China, India, and Brazil would not commit to limiting their own greenhouse gas emissions. A follow-up conference was scheduled for Buenos Aires, Argentina in December 1998. The effectiveness of these conferences remains to be seen, but it's clear that environmental concerns have taken on global, as well as national, importance.

Another significant development: addressing "environmental racism," or the tendency to condone poorer environmental conditions (hazardous waste dumps, illegal/abandoned dumps) in areas populated mainly by people of color. Action has included studies and meetings like the 1991 First National People of Color Environmental Leadership Summit.

Fast Fact

According to the Environmental Protection Agency, the average U.S. resident produces over 0.75 tons of trash a year

STRUCTURE OF THE INDUSTRY

Environmental needs are so pervasive, affecting so many different types of industries, municipalities, and wilderness areas, that the environmental "industry" really is found everywhere. Based upon ECO's general breakdown of environmental jobs, the following is a loose overview of the structure of the industry.

PLANNING

Whether the objective is to save a wildlife habitat, put a transportation system in place, or guide a booming city's expansion, *environmental planners* focus on developing a detailed scheme upfront to help make sure the objective is met. They became very important with the passing of the 1970 National Environmental Policy Act (NEPA) and its rule that any federal project required an environmental impact statement (EIS)—that is, the people carrying out the project had to research and document the effect of the project on the environment. Though most *environmentalists* do some planning, official planners tend to view a situation more widely, address several problems with one plan, and have some political expertise to help get plans approved. They may concentrate on a specific geographic area or a specific environmental issue such as air quality.

Environmental education and communication includes *teachers,* from elementary to college level, *wildlife guides* and *public relations specialists.* All share the task of communicating information about the environment to others. Jobs in this category include everything from *college professors* who instruct tomorrow's scientists and engineers, to *tour guides* who take groups through Yosemite or the Everglades, as well as *corporate communications personnel* who work to put the right "spin" on messages to the public about the company's activities that might impact the environment. Because of their bias, their information may or may not be the most reliable. Currently, an urgent task is to help educate people about landfills and other touchy but critical subjects so that communities can make informed decisions about them.

SOLID WASTE MANAGEMENT

This is the largest of the environmental fields. *Chemists, engineers, recycling coordinators, recycling experts,* and others in this field seek to cut down on the amount and danger of solid wastes—popularly known as garbage. Once, people just put their garbage in dumps; then they started burning it; then they used landfills. Now, *solid waste managers* are trying to reduce the amount of garbage generated in the first place (source reduction), lower the toxicity of garbage that goes into landfills, and find new uses for garbage (such as turning waste into energy). They also seek ways to burn garbage more efficiently, without releasing toxic substances to the air. Individual communities and businesses employ these professionals to help them handle their solid waste; in addition, special companies devoted solely to carting away and disposing of others' solid waste have become a big business.

HAZARDOUS WASTE MANAGEMENT

Hazardous wastes present another, different problem: this is garbage that's potentially lethal to human health or the environment and must be disposed of in special ways. ECO says "millions of tons" of such substances are produced each year. While the chemical industry produces by far the most hazardous waste—up to 70 percent of it—there are many other sources as well, from

THE INDUSTRY, CONTINUED

nuclear reactors to dry cleaners. *Biologists, chemists, engineers, geologists, hydrogeologists,* and many others are employed in this field. Specialists include *radioactive waste managers,* who deal with the various waste materials produced by nuclear energy production and nuclear-powered equipment manufacturing; and *industrial health/hygienists,* who focus on the health effects of exposure to hazardous waste. Opportunities are found with local, state, and federal environmental agencies; on in-house staffs of companies that generate hazardous waste; with consultants or disposal companies; and with emergency response companies, which specialize in dealing with hazardous waste emergencies like chemical spills.

AIR QUALITY MANAGEMENT

Air quality engineers, air quality planners, analytical chemists, and *toxicologists* are just some of the people in this field, which is devoted to the abatement and prevention of air pollution. Under this broad category, people might work on acid rain, ozone depletion, or greenhouse gases. Even more specifically, they might analyze the root causes of these problems. Some might study the effects of sulfur dioxide, nitrogen oxides, hydrocarbons, carbon monoxide, particulates, or mercury, all of which are substances emitted into the air when fuels or trash are burned. Others might study how much sulfur dioxide is emitted in a coal-burning process, or might design, implement, and monitor the effectiveness of scrubbers, which are special devices that clean emissions before they're released into the atmosphere. Still others may monitor the effects of a certain quantity of sulfur dioxide on the human respiratory system, or on animal and plant life.

WATER QUALITY MANAGEMENT

This area focuses on getting polluted water back to the desired quality level—whether for drinking, swimming, fishing, power, irrigation, or other uses. Work involves not only rivers, lakes, canals, and other surface water, but also the water below the ground, known as groundwater. A key area is the recovery and treatment of wastewater for reuse; this is the biggest area of focus within the water quality management category. Other areas of interest are the preservation of wetlands, which are home to certain fish and wildlife, and the reduction of the damaging effects of floods and droughts. *Wetlands ecologists, fish and wildlife scientists,* and *botanists* are just some of the professionals working on wetlands problems.

LAND AND WATER CONSERVATION

In addition to conserving wilderness areas, this category also includes work to ensure better use of land and water for any purpose, so that it can still sustain many different types of plants and animals. There are specific laws for conservation of federally or state-owned land and water, including the National Forest system. Local governments develop their own plans for any land or water in the area not owned by the state or federal government. Special projects in this field include reconstructing destroyed ecosystems. The federal government employs the most people in this category, in such agencies as the Bureau of Land Management, U.S. Fish and Wildlife Service, and National Park Service.

FISHERY AND WILDLIFE MANAGEMENT

This area focuses not only on making sure there are enough of certain species of fish and animals to meet human needs, but also on taking steps to ensure the health of the whole surrounding ecosystem that supports these species. *Wildlife biologists* and *fishery biologists* are key players in this field. Professionals in this category work on wetlands restoration, saving endangered species, cleaning up contaminants, and other projects. Private fisheries and other private companies, plus many U.S. agencies like the Forest Service, use people in this category.

Fast Fact

There are over thirty-seven thousand Superfund sites in the United States.

PARKS AND OUTDOOR RECREATION

Rangers, forest firefighters, geologists, landscape architects, and many others fall into this category. So do *park managers, resource managers, researchers,* and *maintenance personnel.* While the National Park Service uses some of these professionals, it doesn't use a lot. These professionals are found in greater numbers working for other U.S. agencies, or for state, county, or city parks, zoos, and other facilities. One intriguing area in this category is the "re-greening" of city neighborhoods, which involves bringing open park spaces back into urban areas.

FORESTRY

The majority of work in this field involves ensuring healthy forests for use in lumber, paper, and other manufacturing. Smaller percentages work for federal, state, or local governments as *foresters,* helping to conserve and expand forests. An even smaller percentage work for consultants, educational organizations, or nonprofit organizations. Issues in this field include saving endangered

THE INDUSTRY, CONTINUED

species, conserving forest wetlands, and combating pollution. Urban forestry also increasingly is drawing interest; the number of trees within urban areas is dropping, and cities are using foresters to help regain the numbers.

Again, it's important to note that many environmental problems are interrelated and work may crisscross back and forth between these categories. For example, biologists discovered that a major section of the Florida Everglades was contaminated by mercury. After that initial discovery, it took other scientists to figure out how the mercury got there and where it came from. They discovered that incinerated mercury had been carried into the surface water by rain. What started out as solid waste was burned and caused air pollution, which in turn led to the formation of acid rain, which polluted the water in the Everglades.

Federal and state regulatory officials play an important role in the environmental industry. One of their key tasks is to go to the original source or sources of the problem—let's say commercial incinerators—and determine whether the emissions are within legal limits. If not, the offending company might be fined or given a certain amount of time to comply with the regulations. The regulator might return later to determine whether the problem has been fixed. He or she also might eventually testify against the polluter in court.

CAREERS

Many environmental careers are based in the biological sciences, chemistry, or engineering, but others are based in different disciplines. Following are broad definitions of just some of the careers within the environmental industry.

Biologists study all types of living matter, and those specializing in environmental issues focus on how changes in the environment affect living things. Many kinds of biologists work in the field of environmental sciences, including *bacteriologists, biochemists,* and *botanists* such as the *ethnobotanist,* who studies the use of plant life by a particular culture, people, or ethnic group to find medicinal uses of certain plants. Other biologists who specialize in environmental studies include *horticulturists,* who might work on designing safer, less toxic pesticides and herbicides; fishery and wildlife biologists; *microbiologists; mycologists; oceanographers; protozoologists;* and *toxicologists.*

Chemists study the structure and characteristics of natural and artificial substances, and many chemists are involved in environmental research. For example, two chemists at the National Oceanic and Atmospheric Administration discovered that the chemical processes that take place when a volcano erupts have a destructive effect on ozone similar to chlorofluorocarbons. And a professor of chemistry at Harvard led a team of scientists in the studies of damage to the ozone done by aircraft. Chemists, like biologists, usually specialize in a subfield, such as organic chemistry, inorganic chemistry, analytical chemistry, or physical chemistry.

Ecologists study organisms and their place in the environment, and by definition they are usually involved in occupations that promote the protection of the environment. They investigate how environmental damage, such as habitat loss, affects certain species or organisms, and how these organisms interact with each other. They may conduct basic research, work for environmental groups, for the government, or in private industry.

Like other engineers, *environmental engineers* apply the combined principles of mathematics and science to solve problems or create new products. Environmental engineers focus their work on environmental problems. They might help figure out how to clean up an oil spill, for example, or design a process to make coal burn more cleanly. Many environmental engineers work in specific industries or as consultants; others work for the government.

Soil scientists analyze the physical, biological, and chemical characteristics of soils. Although most work is in agricultural applications for the express purpose of maximizing the crop yields of farms, others are needed in conservation applications, such as the study of how deforestation leads to soil erosion. These soil scientists are usually employed by universities.

Fast Fact

According to the Worldwatch Institute and World Resources Institute, worldwide, an estimated sixteen million people have lost their homes and land because of environmental degradation.

Foresters, *aquaculturists* (fish farming specialists), *botanists, geographic information systems (GIS) specialists, land acquisition professionals, land trust or preserve managers,* and *recycling coordinators* are just some of the other careers in this field.

In addition to scientists, many other professionals are involved increasingly in matters relating to the environment. These include *lawyers, consultants, inspectors, planners, writers, editors, lobbyists,* and *politicians.*

EMPLOYMENT OPPORTUNITIES

Employment opportunities are incredibly diverse. Environmentalists can choose to work for federal, state, or local government agencies; on the staff of small, medium, or large private industrial/manufacturing companies; for envi-

Employment Opportunities, continued

ronmental management firms, independent consultants, independent waste disposers/haulers; in research institutions including universities and colleges; at nuclear reactors or other power generators; and so forth. (See the section "Structure of the Industry" for a rundown on some of the specific employment opportunities in the various categories of the environmental industry.) A key trend in government employment: responsibility for environmental action has been steadily shifting to the states, so that more opportunities may now be at that level than was true in the past.

Industry Outlook

Because of growing concern in the United States and the world over for the future health and survival of the planet, most indicators point to a large growth in the field of environmental sciences. Exactly how large is difficult to project because the amount of attention paid to the environment in this country varies with each political administration and other issues, such as the health of the economy and the tension between creating new jobs and protecting the environment. Media attention to one cause or another, such as preserving the wetlands, saving the rain forests, saving the whales, or recycling, waxes and wanes, but helps to periodically remind people that significant environmental problems are here and need answers right now.

Another factor in favor of environmental cleanup is the waning of the nuclear arms race. Industrialized nations now have more resources to find alternatives to fossil fuels, protect the ozone layer, put a stop to habitat and species destruction, and develop methods for conserving water, energy, and other resources.

In the United States, spending on the environment has risen from $30 billion in 1972 to an estimated $295 billion in 1998. The environmental industry as a whole is growing more slowly now than in the past—from 16 to 30 percent per year in the mid- to late 1980s to about 2 percent in 1991, 4 percent in 1992, and a projected 5 to 7 percent for the next 7 years. *Environmental Business Journal* says there were 793,159 environmental jobs in 1988, 1,073,397 in 1992, and 1,327,150 in 1997.

Section 2
WHAT DO I NEED
TO KNOW ABOUT
Careers
?
environment

Ecologist

Summary

Definition
Ecologists examine the complex relationships between living organisms and the physical environment. They combine biology, which includes the study of both plants and animals, with physical sciences, such as geology and geography.

Alternative Job Titles
Botanist
Environmental scientist
Zoologist

Salary Range
$10,000 to $45,000 to $85,000

Educational Requirements
Bachelor's degree
Master's or Ph.D. recommended

Certification or Licensing
None

Employment Outlook
Faster than the average

High School Subjects
Biology
Chemistry
Computer science
Earth science
English (writing/literature)
Geology
Mathematics

Personal Interests
Botany
The Environment
Science
Zoology

In the filtered light of the forest, Professor Tim Schowalter and Mark, a graduate student, hike along a crudely cut trail. They pause briefly at a predetermined point. Tim fights his way through dense undergrowth to a tree growing several yards from the trail. Using a long, closeable net lined with a plastic bag, he is able to snap an upper branch from the thirty-foot tree. Tim returns to the trail with the sample, and quickly uses the drawstring on the net to seal the sample. Later, Tim and his assistant will examine the sample to identify and record the organisms living or resting on the branch.

"A sample is like a snapshot," Tim explains. "It gives you a good idea of what's living on or associated with the plant."

A professor of entomology at Oregon University, Tim is conducting ecological research in the Luquillo Experimental Forest in Puerto Rico. He is studying the recovery of the tropical forest in the wake of Hurricane Hugo. He is also examining the impact of the hurricane on species diversity.

As the two continue down the trail to the next sample point, Mark points, with a mixture of wonder and trepidation, at a flying walking stick. Tim grins. "You know," he observes, "the first time I came down here I was really impressed by the dramatic fauna you can see in this tropical forest. But the truth is, we have the same amazing diversity of species back in Oregon. We've gotten accustomed to seeing the plants, insects, and other arthropods that live in Oregon, though, so we don't always stop to appreciate them. One of the wonderful things about the study of ecology is that it really reminds us to notice the amazing intricacies of the world."

When Tim and Mark return to camp after several hours of collecting samples, they learn that the electricity is not working. "Well," says Tim with a laugh, "it's not the first time I've had to sort samples by the light of a lantern."

What Does an Ecologist Do?

Ecologists study the relationships between living organisms and their environment. They try to understand the way changes in the environment affect living organisms. An ecologist might, for example, study the effect of pollutants on species diversity within a river. Another ecologist might explore the impact logging practices have on arthropods and plant life within a forest.

Because the connections between living organisms and the environment are so diverse and intricate, most ecologists concentrate on studying one ecosystem or many ecosystems that share similar characteristics. An ecosystem is a single community of organisms that interacts with a specific environment. Physical characteristics, such as climate, altitude, and topography, define an ecosystem's environment. Coniferous forests, rain forests, rivers, savannas, and tundras are all different types of ecosystems.

Because the relationships within an ecosystem are extremely complex and delicate, even small environmental changes can upset the delicate balance within the ecosystem. The survival of each species within an ecosystem is dependent, to some degree, on the survival of every other species within that ecosystem.

Lingo to Learn

Arthropod: *An animal with an exoskeleton, a segmented body, and jointed appendages.*

Canopy: *The upper layer of a forest, created by the foliage and branches of the tallest trees.*

Coniferous: *A coniferous forest is composed of trees that bear cones.*

Ecosystem: *A community of animals and plants and their interaction with the environment.*

Effluent: *Wastewater or sewage that flows into a river, lake, or ocean.*

Entomology: *The study of insects.*

Invertebrate: *An organism that does not have a backbone.*

Riparian zone: *Forest or grass growing on the banks of a stream. The riparian zone can prevent soil erosion.*

Savanna: *A flat, grassy plain found in tropical areas.*

Tundra: *A cold region where the soil under the surface of the ground is permanently frozen.*

Watershed: *The gathering ground of a river system, a ridge that separates two river basins, or an area of land that slopes into a river or lake.*

Each organism plays a vital role in the food chain. Green plants "fix" energy through photosynthesis. That is, they capture solar energy in the chemical bonds of carbohydrates synthesized from water and carbon dioxide. Some insects and animals obtain that energy by eating the plants. Others obtain energy by eating insects or animals that have consumed plants. If one species fails, the organisms that feed on that species may, in turn, become endangered. Living organisms also release chemicals into the atmosphere, water, and soil (depending on where they live) as they fix, consume, or process energy. Each of these chemicals plays an important role in sustaining life within an ecosystem.

Many ecologists devote their careers to studying the forces that can upset ecosystems. They attempt to find ways to prevent disruption from occurring. If ecosystems already have been damaged, ecologists look for ways to help them recover.

In order to understand any ecosystem, ecologists must consider many factors. To understand events within a single river ecosystem, for example, ecologists must study the behavior of the living organisms within the river and look for evidence of disease or pollutants within the organisms' cells. They must evaluate the quality of the water in the river. They must study the river's banks for traces of soil erosion. They also must consider the slope of the river, the proximity of any heavy industry or sewage treatment plants, and local farming practices. To understand just one ecosystem, ecologists must combine many different areas of knowledge, including zoology, cellular biology, geography, and geology.

Ecologists gather information in many ways. They usually collect samples from the ecosystem or ecosystems they are studying. Using nets lined with plastic bags, they collect samples of plant life and the invertebrates and other organisms that dwell on or amidst plant life. Small containers are used to gather soil and water samples. In addition to collecting samples, ecologists rely on satellite data about an ecosystem's geography. They compare data collected from one ecosystem to immense databases of information about comparable ecosystems. This comparison enables ecologists to determine whether an ecosystem is deviating from normal standards.

Once an ecologist has gathered significant data about an ecosystem, he or she must interpret the data, which can be a painstaking process. Ecologists draw conclusions about measurable changes within an ecosystem, about their causes, and about their possible long-term consequences. Ecologists also make recommendations for protecting or restoring ecosystems.

What Is It Like to Be an Ecologist?

Although ecologists study ecosystems, most spend the majority of their time indoors, working in laboratories, offices, or classrooms. Tim Schowalter estimates that he spends 10 percent of his time teaching and advising graduate students and another 10 percent in university service. Tim serves on the faculty senate, several departmental committees, the budget committee, the advisory committee, and the selection committee. He spends another 10 percent of his time serving on Ecological Society of America committees and national and state policy advisory committees on forest health and ecosystem management. His remaining time is devoted to research. Tim is quick to note, however, that he conducts only a fraction of his research in the field. He spends most of his time in his office analyzing data, or in laboratories sorting samples.

Susan Cormier, a branch chief at the Environmental Protection Agency's (EPA) Office of Research and Development, spends very little time in the field. "I think I got out in the field three days this year, and that was only because I decided to really treat myself," she comments, somewhat ruefully.

As a branch chief, Susan plans and interprets major research projects. Her department concentrates on conducting vulnerability assessments of stream ecosystems. To perform these assessments she works to develop methods to measure the conditions of stream ecosystems, as well as diagnose the stressors and their sources. Susan identifies the questions that each research project should answer and then develops strategies for finding answers to the questions. Her staff helps her implement the research strategies.

Susan's research team examines the diversity, abundance, and spatial distribution of fish and invertebrates in various streams. They measure the sediment load in water and note the presence, or lack, of riparian zones along rivers' banks. Riparian zones are grassy or forested areas that separate rivers from farms, roads, or houses. Riparian zones minimize soil erosion and act as buffers between the river ecosystems and the stressors and pollutants caused by development. Finally, Susan and her staff compare the information gathered from one stream to information gathered from many similar streams. This comparison enables them to assess the stream's deviation from normal standards.

Using information gathered through all this research, Susan assesses the health of an ecosystem. After considering all the factors involved, she develops a prognosis for the ecosystem's long-term health and provides recommendations for improving or protecting the ecosystem.

Susan delivers these vulnerability assessments to the scientists, politicians, and activists in communities that surround the ecosystems in question.

"We are trying to do community-based ecosystem protection," she explains. "Our role is to educate the public and to give counties and states information that they can use for planning practices. Our reports also go to Congress, to give representatives a good picture of the health of our nation's ecosystems."

"I feel that I'm doing something that matters. When I tell my kids what I do, they understand that Mommy is trying to protect the earth. That gives me a good feeling."

In addition to planning and analyzing research projects, Susan is responsible for drafting reports and managing a team of scientists. She also must perform a variety of administrative responsibilities, including reviewing timecards, conducting performance reviews, managing budgets, and procuring resources.

"There are jobs I could get where I could do more actual field research," Susan observes, "but I can make a bigger impact where I am. I feel that I'm doing something that matters. When I tell my kids what I do, they understand that Mommy is trying to protect the earth. That gives me a good feeling."

Have I Got What It Takes to Be an Ecologist?

Ecologists, like other scientists, must be intelligent and possess intellectual curiosity and must be able to think both analytically and creatively about complex issues. Most importantly, they must be excited about understanding the environment and committed to preserving it.

To be a successful ecologist, you should:

Be intelligent and possess intellectual curiosity

Be excited about the environment and committed to preserving it

Have good speaking and writing skills

Be willing to work closely with others and share ideas

To the surprise of many would-be ecologists, communication skills are every bit as important as the ability to take accurate measurements and conduct good research. Ecologists must be able to communicate ideas to other scientists, to regulatory agencies, and to the public. They must, therefore, be able to speak and write clearly. According the *Princeton*

Have I Got What It Takes?, continued

Review, most ecologists consider writing skills the second or third most important skill for succeeding in this field.

Most science is collaborative. Ecologists must be able to work closely with others in their field and should enjoy sharing ideas and being challenged by others' questions.

How Do I Become an Ecologist?

Because ecology requires a multidisciplinary approach, many ecologists actually study other disciplines before embarking on a career in ecology. This is true of both Tim Schowalter, who earned a doctoral degree in entomology, and of Susan Cormier, who earned a doctoral degree in cellular biology.

According to a 1992 survey conducted by the Ecological Society of America, more than 27 percent of ecologists earned their highest degree in biology. Another 19 percent earned their degrees in zoology and nearly 14 percent earned their degrees in botany. Ecologists who earned their highest degrees in ecology were, in fact, a distinct minority.

"We hire people who know ecology, but who also know statistics, geography, or genetics."

"To be successful in ecology, you have to be multidisciplinary in approach," comments Susan Cormier. "If you can bring more than one area of knowledge to the table you will be better off. We hire people who know ecology, but who also know statistics, geography, or genetics.

"Ecologists who do not have a well-rounded science background usually will end up in low-paying, uninteresting jobs," she adds.

Education

High School

High school students who are interested in becoming ecologists should concentrate on science and math classes. They should not, however, neglect other disciplines. English, for one, can provide students with useful experience in writing well and easily.

"It is critical for an ecologist to be able to write and read well," notes Susan Cormier, "so students should take courses that will polish those skills. The biggest problem I see with ecologists at the professional level is that, after conducting their research, they get stuck at the writing stage. Because I've seen so many people with this problem, I advise students to concentrate on developing writing skills."

Postsecondary Training

Once in college, students should continue to study science. Ecology courses are important, but students should take biology, chemistry, meteorology, and zoology courses, as well. Geography and geology can be equally helpful in preparing a student for a career as an ecologist. Because ecologists amass and analyze immense amounts of data, students should also take math, computer, and statistics courses.

While people with undergraduate degrees in ecology can find employment as laboratory technicians or field researchers, the vast majority of ecologists have master's or doctoral degrees. "If you want to be in a decision-making position, you pretty much have to have an advanced degree," says Tim Schowalter. "Most ecologists who are conducting their own research have Ph.D. degrees."

Many colleges and universities throughout the United States offer doctoral degrees in ecology or related fields. According to the Ecological Society of America, a great many practicing ecologists received doctoral degrees from the University of California, Berkeley; the University of Wisconsin, Madison; Cornell University; the University of Washington; and Duke University. In Canada, many ecologists received doctoral degrees from the University of British Columbia, the University of Alberta, and the University of Toronto.

Internships and Volunteerships

Students also should strive to gain as much practical experience as possible by volunteering for environmental organizations or by helping an ecologist conduct research. Research positions for high school students are not abundant, but some ecologists will hire students to collect and sort samples. Tim Schowalter, for instance, usually hires several high school students each summer.

"I hire both college and high school students," Tim explains. "These students learn to sort and weigh samples. They also learn to use different software programs for recording information. I try to get each student out in the field once or twice, to get them excited. I also give each student an overview of

HOW DO I BECOME AN ECOLOGIST?, CONTINUED

the proposal and the methods involved, so that they can see the significance of the study."

Though he tries to give each student some field experience, Tim concedes that the students he hires spend the majority of their time measuring and sorting data. He believes this is a fairly accurate reflection of a scientist's life. "Science is mostly grunt work," he explains, "and only about 5 percent 'Eureka!' factor."

WHO WILL HIRE ME?

According to the Ecological Society of America, universities and four-year colleges employ about 60 percent of the ecologists currently practicing. The national government is the second-largest employer category, employing about 12 percent of practicing ecologists. Within the national government, both the Environmental Protection Agency and the National Park Service are major employers of ecologists.

Though the salaries for ecologists in academia and government are roughly comparable, the positions are quite different. Susan Cormier observes, "It's definitely a trade-off. Academia has a lot of pluses. Academics deal with less bureaucracy. They deal with a wider variety of people, which can be very

Global Warming

Ecologists today are struggling to predict the possible consequences of a phenomenon called global warming. *Global warming is the slow rise in our planet's average temperature. The earth's average temperature has increased 0.5 to 1°F over the last century and scientists estimate that it may increase by 1.8 to 6.3°F in the coming century.*

Global warming has the potential to significantly change the climate in most regions of the world and to alter the balance within many of our planet's ecosystems. It may cause flooding in some regions and drought in others. It could lead to the spread of diseases usually associated with warm climates, such as malaria. Ecologists and climate scientists from around the globe are concerned about how this change in our environment will affect humans and other living organisms.

What causes global warming?

Global warming is caused by an increase in the so-called 'greenhouse gases,' such as carbon dioxide, methane, and nitrous oxide, that trap heat within the earth's atmosphere, much as glass panels trap heat within a greenhouse.

Greenhouse gases are not all harmful, however. Without them, we could not survive on the earth's surface. The sun heats the earth, and the earth radiates this heat back into space. If atmospheric greenhouse gases did not trap some of this heat within the earth's atmosphere, our planet would be extremely cold and uninhabitable.

stimulating. They also have much more flexibility. If they prefer to do their writing at home, where they have fewer distractions, they can. In government jobs, the hours are pretty rigid. We're not even supposed to work late. That 'punch the clock' mentality can really get in the way."

"On the other hand," she continues, "ecologists who work for universities have to be very competitive to get grant money and to find summer projects that will augment their nine-month academic salary. As a government employee, I have an annual salary."

Smaller percentages of ecologists work for consulting firms, state or local governments, or environmental nonprofit organizations. A small number of ecologists work for private industry. Ecologists who work for private industry help companies achieve their business objectives in ways that are least disruptive to surrounding ecosystems. They also help comply with environmental regulations. Most companies employ only one or two ecologists, though, so these positions can be difficult to find.

Where Can I Go from Here?

Opportunities for advancement are limited for individuals who have only a bachelor's or master's degree. Those who earn doctoral degrees, however, may

---

Global warming occurs when human beings add to our atmosphere's greenhouse gases by driving cars and trucks, heating buildings, and running factories. The United States is one of the biggest contributors to this problem. In 1997, the United States contributed 30 percent of the total global greenhouse gas emissions.

What Can We Do?

Governments around the world are considering regulations that will limit the greenhouse gases that businesses and individuals create. In the meantime, we can all take steps to minimize the greenhouse gases we add to the atmosphere. Here are a few of things you can do to help:

Whenever possible, walk, bike, or use public transportation instead of driving.

Insulate your walls and ceilings.

Wash clothes in warm or cold water instead of hot water.

Use the energy-saving settings to dry dishes in your dishwasher.

Turn down your water heater thermostat to 120 degrees and wrap your water heater in an insulating jacket.

Clean or replace your air filters.

Reduce the waste you create by choosing reusable products and products with less packaging.

Recycle.

WHERE CAN I GO FROM HERE?, CONTINUED

advance in many ways. Ecologists who conduct research can gain recognition for their work by publishing reports and articles in scientific journals. Highly visible ecologists may accept public speaking engagements, which can significantly augment their income. Those who work for universities or colleges also can advance by serving on committees or assuming administrative responsibilities. Some ecologists become department chairs. These positions require considerable administrative work but offer higher salaries. Ecologists who work for smaller academic institutions also can advance by seeking positions within larger or more prestigious institutions.

Within government positions, ecologists may advance by assuming positions of greater responsibility, such as supervisory or branch chief positions. In these positions individuals must manage staff personnel and other scientists. Supervisory positions usually entail significant administrative work. Individuals in these positions may find themselves spending more time hiring and managing employees, conducting performance reviews, and completing the paperwork than they devote to actual science.

Senior ecologists within the government are, however, able to influence legislation, inform government officials, and educate the public. Susan Cormier comments, "It is very rewarding for me to see people actually use my information and ideas to protect the environment."

As in government and academia, ecologists in industry advance by assuming more responsibility. Most large companies employ very few ecologists, however, so management opportunities may be quite limited. After gaining experience by working within private industry for a length of time, some ecologists choose to become consultants. Ecologists who are able to develop a large and varied clientele can significantly increase their income and gain prominence in the field.

Related Jobs

Biochemists
Biologists
Botanists
Geographers
Geologists
Marine biologists
Meteorologists
Microbiologists
Physiologists
Zoologists

WHAT ARE SOME RELATED JOBS?

The U.S. Department of Labor classifies ecologists with biological and medical scientists (DOT). This category includes biologists, marine biologists, biochemists, botanists, microbiologists, physiologists, and zoologists. Because ecology requires a multidisciplinary approach, it is also related to science disciplines that study the earth, climate, atmosphere, and the chemical composition of our environment. Related jobs also include geographers, geologists, and meteorologists.

What Are the Salary Ranges?

Ecologists' salaries vary greatly depending on the individual's level of education, place of employment, and years of experience. A few generalizations are possible, however. Ecologists who have doctoral degrees typically earn more than those who have less education. Ecologists who work for private industry usually earn higher salaries than those in academia or governmental positions. Those who work for nonprofit environmental agencies earn the lowest salaries of all.

According to the Ecological Society of America, salaries for ecologists can range from $10,000 per year to $175,000 and above. Salaries of $175,000 and above are quite rare, however—most experienced and successful ecologists in academia and government can hope to earn salaries of no more than $85,000. The median income for ecologists is about $45,000.

What Is the Job Outlook?

Ecological careers are expected to increase at a faster rate than the average for all occupations through 2006. Environmental concerns are fueling the surge in this field, as nations around the world become more aware of the dangers posed by pollutants, pesticides, greenhouse gasses, and uninhibited population growth. People who have the skills to help communities and countries find practical ways to protect or repair ecosystems will find ecology a challenging and rewarding profession.

Environmental Lobbyist

SUMMARY

DEFINITION

Lobbyists *try to influence legislators to support legislation that favors certain causes or special interest groups. Environmental lobbyists encourage legislators to support bills that will protect our environment.*

ALTERNATIVE JOB TITLES

Lobbyist

SALARY RANGE

$12,000 to $30,000 to $80,000

EDUCATIONAL REQUIREMENTS

Bachelor's degree

CERTIFICATION OR LICENSING

None

EMPLOYMENT OUTLOOK

Faster than the average

HIGH SCHOOL SUBJECTS

Biology
Ecology
English (writing/literature)
Government
History
Speech

PERSONAL INTERESTS

Current events
Economics
The Environment
Law
Politics

On the day the House of Representatives is scheduled to vote on HR1127, the National Monument Fairness Act, Cindy Shogan arrives at the office early in the morning. She knows she's in for a chaotic day.

Despite its innocuous-sounding name, Cindy, who is an environmental lobbyist for the Southern Utah Wilderness Alliance, considers the National Monument Fairness Act a dire threat to conservation efforts. The bill is intended to weaken the Antiquities Act of 1906, which gave presidents the power to protect natural and cultural resources by creating national monuments. Cindy believes that the Antiquities Act of 1906 is among the most important conservation laws ever passed. To protect it, she has joined a coalition of environmental organizations to fight the passage of the National Monument Fairness Act.

Cindy begins her marathon day with a conference call. She and her colleagues quickly map out a strategy for convincing members of the House that this bill will torpedo efforts to protect significant public lands from harm. Their first task is to deliver "Dear Colleague" letters to all 435 members of the House. Cindy races over to Capitol Hill, known among lobbyists and other Washington insiders as "the Hill," to deliver her share of the leaflets.

When she returns to her office just after 12:30, Cindy finds that she has received a shipment of three-inch replicas of the Statue of Liberty. She and a colleague frantically tie small tags to the miniature statues. The tags read, "Don't torch our national heritage. Vote 'no' on HR1127." Just as soon as the tags are tied, Cindy and several other lobbyists return to the Hill to deliver the tiny statues to each of the 435 members of the House of Representatives.

At 2:00, Cindy and other members of the conservation coalition stage a press conference. Standing before enormous photographs of the Grand Canyon and other national monuments, a sympathetic member of Congress delivers a stirring criticism of the National Monument Fairness Act. As he speaks, members of the conservation coalition use white paint to make the national monuments 'disappear.' "We have to do something sort of dramatic and symbolic to get attention," Cindy explains.

Cindy's day is just getting started.

What Does an Environmental Lobbyist Do?

Lobbyists are people who strive to influence legislation on behalf of a special interest group or a client. Like other lobbyists, environmental lobbyists strive to influence state or federal legislation in order to achieve a goal or to benefit a special interest group. Environmental lobbyists, however, deal specifically with environmental issues. Most environmental lobbyists work for environmental protection organizations, such as the Natural Wildlife Federation, Sierra Club, or the Natural Resource Defense Fund. They urge legislators and other government officials to support measures that will protect endangered species, limit the exploitation of natural resources, and impose stricter antipollution regulations.

Occasionally looked upon as mere influence peddlers, lobbyists actually serve an important role in the democratic process. Government officials and legislators must understand and make decisions about innumerable issues. They cannot possibly be experts in every area. Consequently, many rely upon lobbyists to provide them with information about important issues. Environmental lobbyists compile information about the probable impact of various measures on the environment and are sometimes invited by legislators to help them draft new bills.

Lingo to Learn

Act: *A bill that has been passed by Congress. If signed by the President, the act becomes law.*

Bill: *A written plan for a new law, which must be discussed and voted upon by Congress.*

Constituent: *A voter represented by an elected official.*

Veto: *To stop a bill from becoming a law.*

Environmental lobbyists strive to influence legislators and government officials through both direct and indirect lobbying. Direct lobbying involves reaching legislators themselves. Environmental lobbyists meet with members of Congress, their staff members, and other members of government. They call governmental officials to discuss the impact various measures might have on the environment. They sometimes testify before congressional committees or state legislatures. They distribute letters and fact sheets to legislators' offices. They sometimes try to approach legislators as they travel to and from their offices, and some lobbyists ask legislators who share their views to broach issues with other, less sympathetic legislators.

In another form of direct lobbying, environmental lobbyists strive to persuade members of Congress to serve as cosponsors for bills the lobbyists support. When a member of Congress becomes a cosponsor of a bill, his or her name is added to the list of members supporting that measure. Lobbyists typically assume that cosponsors will vote to support the bill. They also use the list of cosponsors to influence other members of Congress to support a measure. A bill's chances of one day becoming a law dramatically improve as more members agree to serve as cosponsors.

Indirect lobbying, also called grassroots lobbying, involves educating and motivating the public. The goal of indirect lobbying is to encourage members of the public to urge their representatives to vote for or against certain legislation. Environmental lobbyists use an array of indirect techniques. They issue press releases about pending legislation, hoping to inspire members of the media to write topical articles. They mail letters to citizens, urging them to write or call their representatives. They post information on the Internet and sometimes go door-to-door with information to mobilize members of environmental groups. On rare occasions, they take concerned citizens to state capitals or to Washington, DC, to meet with representatives.

For both direct and indirect lobbying efforts, environmental lobbyists try to form coalitions with other environmental groups. Members of these coalitions work together because they have a common interest in protecting the environment. By pooling information and resources, members of the coalition can be more effective in reaching the public and members of government.

Some environmental lobbyists also support political candidates who are likely to support measures that protect the environment. They promote these candidates by distributing positive information to the public and by raising money for their campaigns.

What Is It Like to Be an Environmental Lobbyist?

Kevin Kirchner, an environmental lobbyist for Earthjustice Legal Defense Fund compares his job to that of Sisyphus, the king in Greek mythology who was condemned to push a boulder up a hill, over and over again, for all eternity.

"Every time he got the boulder up the hill, it would roll back down again," says Kevin. "My job is a lot like that. We constantly are fighting against special interest groups that have massive war chests. They can afford to provide huge campaign donations and pay for TV commercials. As a not-for-profit group, we aren't allowed to contribute to campaigns and we rarely have the money to do any advertising."

The Antiquities Act Battle

Lobbyists must grapple with extremely complex issues. The National Monument Fairness Act (HR 1127) is no exception.

If passed, the National Monument Fairness Act will curtail presidents' power to create national monuments. If it becomes law, presidents will need congressional approval to create national monuments larger than fifty thousand acres. The National Monument Fairness Act is intended to weaken the Antiquities Act of 1906, which gave presidents the ability to create national monuments without congressional approval.

Representative Jim Hansen, of Utah, proposed the National Monument Fairness Act after President Clinton created the Grand Staircase-Escalante National Monument in Utah. This new monument, which the President created without consulting any government officials or representatives from Utah, occupies 1.7 million acres of Utah land.

According to Rep. Hansen, the Antiquities Act of 1906 was created at a time when there were no laws in place to protect archeological sites. The Act, he says, was intended to enable presidents to move quickly to protect objects of historical interest by creating national monuments on small tracts of land. Since 1906, though, Congress has passed many laws designed to protect large areas of land. Rep. Hansen argues that, by creating a huge national monument without congressional approval, President Clinton was abusing the Antiquities Act and circumventing the democratic process.

Conservationists, on the other hand, believe that the Antiquities Act of 1906 is one of the most important conservation laws ever enacted by Congress. Since 1906, they argue, the Act has been used to proclaim 105 national monuments that protect outstanding lands from such threats as mining, logging, oil drilling, and uncontrolled development. The Act has ensured the protection of the Grand Canyon, Denali, the Petrified Forest, Glacier Bay, Acadia, and other American treasures. Conservationists believe that, if the National Monument Fairness Act becomes law, our nation's ability to protect other significant public lands from harmful uses will be severely hampered.

On October 7, 1997, the House of Representatives voted in favor of the National Monument Fairness Act. The bill will now go the Senate, where it will be reviewed and put to a vote.

If students are interested in obtaining more information about the Antiquities Act of 1906 or the National Monument Fairness Act, they should write, call, or send email messages to their representatives in Congress.

According to Cindy Shogan, lobbyists' schedules are determined by Congress' schedule. "If Congress is on break," she says, "people dress more casually and spend time catching up on research, paperwork, and grassroots lobbying. When Congress is in session, you're in constant crisis mode."

"Crisis mode," for an environmental lobbyist, entails twelve- and fourteen-hour days, irregular hours, and frequent trips to Capitol Hill. Lobbyists usually are involved in several lobbying campaigns simultaneously. Whenever Congress is considering any of the issues a lobbyist is promoting or opposing, lobbyists must take every opportunity to meet members of Congress, committee members, and congressional aides. Environmental lobbyists also engage in frequent strategy sessions with colleagues in the environmental movement. They distribute fact sheets and "Dear Colleague" letters on the Hill. They send faxes, issue press releases, and mail information.

Because most environmental lobbyists work for not-for-profit organizations, they often have limited staff and even more limited budgets. Consequently, environmental lobbyists usually combine highly professional skills, such as scientific or legal expertise, with clerical capabilities. In other words, environmental lobbyists must be willing to stuff envelopes as well as meet with senators.

"You feel like you're beating your head against the wall a lot, but it makes the occasional wins that much more rewarding. You know you're accomplishing something that will last for generations."

Environmental lobbyists often face frustrating setbacks. Measures they support can take years to wend their way through Congress. Along the way, they can be altered and weakened almost beyond recognition. "You feel like you're beating your head against the wall a lot," says Kevin, "but it makes the occasional wins that much more rewarding. You know you're accomplishing something that will last for generations."

HAVE I GOT WHAT IT TAKES TO BE A LOBBYIST?

Environmental lobbyists must be tenacious, self-motivated individuals. They must have excellent communications skills. They must be able to work well in teams and perform well under pressure. They must understand the political process. Because lobbyists must be able to approach governmental officials and powerful legislators, they should be confident, outgoing individuals. Most importantly, environmental lobbyists must be committed to protecting the environment.

To be a successful environmental lobbyist, you should:

Be confident, tenacious, and self-motivated

Have excellent communication and persuasive skills

Know how our political system works

Be able to work as part of a team

Be committed to protecting our environment

Be able to think quickly on your feet

Have an ability to handle stress and long work hours

"Environmental lobbyists," says Cindy, "have to be able to think quickly on their feet. If one approach isn't working, they have to be able to shift gears. They have to be flexible. They also have to have a keen strategic sense. They have to be able to look at all the angles and really think through a course of action."

According to Kevin, a sense of humor can be a vital tool for an environmental lobbyist. "This is high stress work," he explains. "It's easy to get frustrated. You've got to be able to find humor in what's going on. I also think environmental lobbyists must have an unwavering dedication to accuracy and the truth. Legislators will only listen to lobbyists who have credibility."

HOW DO I BECOME AN ENVIRONMENTAL LOBBYIST?

No one academic path leads directly to a lobbying career. Most lobbyists come to the profession from other disciplines and other jobs. Some have political experience, others have scientific, economic, or legal backgrounds. This previous experience can be extremely useful to environmental lobbyists, who must be able to assess the environmental and economic impact and to identify the legal strengths or weaknesses of various measures.

EDUCATION

High School

Students interested in becoming environmental lobbyists should pursue a well-rounded education. They should, of course, study civics and history to gain an understanding of our country's political system. They also should take biology, ecology, and chemistry in order to learn about the scientific issues behind environmental legislation.

In addition to understanding politics and science, lobbyists must have a number of very practical skills, including the ability to use computers and the ability to write and speak clearly. Students should, therefore, take computer courses, speech classes, and English.

Postsecondary Training

While there are no specific requirements for environmental lobbyists, most have college degrees; a growing number also have advanced degrees.

During their undergraduate studies, students should continue to take courses that will help them understand the complex issues behind legislation and gain the practical skills that will make them effective lobbyists. Students should take courses in environmental science, geography, and geology. They should also study political science and history, which will help them understand how our political system developed and help them prepare to function within that system. Students who hope to become environmental lobbyists should also study economics, because lobbyists must be able to assess the probable economic impact of pending legislation.

Lobbyists must be able to do more than understand the issues, however—they must also be able to write and speak about them. They must be able to influence the way other people think about issues. Communication, public relations, and English all can be helpful courses for the future environmental lobbyist.

Students who choose to pursue advanced degrees will find that having special areas of expertise, such as ecology, economics, or law, coupled with broad undergraduate backgrounds, will help them find interesting positions.

INTERNSHIPS AND VOLUNTEERSHIPS

Students who are interested in this career should also consider serving as interns for environmental organizations. Some colleges and universities will award academic credits for internship experiences. Internships can also help students gain hands-on experience, learn about the issues, and meet potential employers. Serving as interns, says Cindy Shogan, "can help students decide what they want to do within the environmental movement. It gives them a better idea of the options. It also helps them learn about other environmental organizations, so they can decide which ones appeal most to them philosophically."

How Do I Become an . . . ?, continued

Political or governmental experience is also invaluable for would-be lobbyists. Students should consider seeking staff positions within legislators' offices or pursuing governmental internships. Several agencies in Washington, DC, offer governmental internships. Among these are the Library of Congress, the U.S. Department of Agriculture, and the School of International Service at American University.

Students also can gain valuable practical experience by volunteering for environmental organizations. This practical experience can help students understand the issues and obstacles that environmental lobbyists encounter. By volunteering to work for local political campaigns or by serving as pages in Congress, students can learn about our country's political system.

Certification or Licensing

Though there is no certification process for lobbyists, federal law does require lobbyists to register if they lobby among federal agencies or bodies. Many states also regulate lobbyists' activities. Some require lobbyists to file reports outlining their activities. For additional information about the registration requirements for lobbyists, students should contact their state governments or the American League of Lobbyists.

Who Will Hire Me?

Many of our country's most respected environmental protection organizations employ environmental lobbyists. The Nature Conservancy, Sierra Club, Wildlife Federation, Wilderness Society, and Friends of the Earth are just a few of the organizations that are actively involved in lobbying on behalf of our environment. Students interested in becoming environmental lobbyists should contact various environmental organizations to discuss lobbying activities.

Where Can I Go from Here?

Environmental lobbyists can advance by gaining experience, demonstrating their abilities, or earning advanced degrees. In large environmental organizations, they also may encounter opportunities to assume management responsibilities.

Unlike lobbyists who work for special interest groups that represent major industries, environmental lobbyists are rarely motivated by ambition.

Most choose the profession out of a genuine desire to protect our country's natural resources.

Related Jobs

Campaign managers
Ecologists
Economists
Fundraising directors
Lawyers
Political scientists
Public relations representatives
Public service directors
Sales or service promoters
Writers

What Are Some Related Jobs?

The Department of Labor classifies lobbyists with public relations management occupations (DOT) and other workers in public relations (GOE). These categories include fundraising directors, public relations representatives, and sales or service promoters, employee relations representatives, and public service directors.

Lobbying is an unusual profession, however, in that it draws from the expertise of many other professions. This is particularly true of environmental lobbying. Lawyers, ecologists, economists, political scientists, writers, and campaign managers all have skills useful to environmental lobbying.

What Are the Salary Ranges?

Environmental lobbyists usually work for not-for-profit organizations with extremely limited budgets. Consequently, their salaries tend to be much lower than those of other lobbyists. While most lobbyists may earn anywhere from $12,000 to $700,000, depending on the groups they represent and their years of experience, environmental lobbyists are more likely to earn between $12,000 and $80,000.

"I would be shocked to learn that any of my colleagues were earning more than $80,000."

"I would be shocked," says Cindy Shogan, "to learn that any of my colleagues were earning more than $80,000."

Kevin Kirchner comments, "Environmental lobbyists must have an enormous commitment on a personal level because they just don't get paid as much as they could in other jobs and they often work longer hours."

What Is the Job Outlook?

Regrettably, there is no shortage of environmental concerns in our country. As long as people continue to pollute our air and water, cut down forests, develop land, and mine the earth, environmental groups will continue to fight for legislation that will protect our natural resources. This profession is, therefore, expected to grow about as fast as the average until 2006.

The nation's economy can affect environmental protection organizations, which are largely funded by donations. During recessions, people may not be able to give as generously to not-for-profit organizations. Environmental protection organizations may, in turn, be forced to cut back on their lobbying efforts.

Groundwater Professional

Summary

Definition
Groundwater professionals study water supplies, test the quantity, quality, and availability of underground water. The also strive to find ways to protect and preserve water resources.

Alternative Job Titles
Environmental engineer
Groundwater geologist
Groundwater scientist
Hydrogeologist
Hydrologist
Water chemist

Salary Range
$25,000 to $30,000 to $60,000

Educational Requirements
Bachelor's degree

Certification or Licensing
Voluntary

Employment Outlook
About as fast as the average

High School Subjects
Biology
Chemistry
Computers
Earth science
English (writing/literature)
Geology
Mathematics

Personal Interests
The Environment
Science
Travel

It's a bright September morning in St. Paul, Minnesota. Frost coats the windshield of Jim Lundy's truck, but it's still quite early. Jim guesses the thermometer will reach seventy by noon. Quickly rifling through his briefcase, he reassures himself that he has all the necessary papers. Jim tosses a bag of bagels in his truck and is ready to begin the six-hour drive.

Jim is investigating groundwater contamination near a small town in northern Minnesota. He has carefully examined all the available information, but needs to collect soil and water samples to answer crucial questions about the nature and extent of the contamination, its proximity to the town's water supply, and the best means for cleaning it up. Before visiting the site, Jim has carefully compiled a list of the questions that must be answered and has mapped out a strategy for collecting the appropriate samples.

By 2:00 PM, Jim has reached his destination. He greets Joe Schmidt, the drill rig operator, and the two quickly get down to business. Jim marks six spots within a six-hundred-foot radius. He asks Joe to drill a twenty-five-foot sampling well at each point. As Joe is drilling the second well, he encounters a problem with the drill rig. While Joe heads into town to see if he can find someone to work on the rig, Jim calmly continues to collect samples from the first well. "Something always goes wrong during an investigation," he comments with a shrug. "We're lucky to be near town on this one. It's tougher when you're out in the woods with no one around."

In less than an hour, Joe returns with surprisingly good news. "I found a guy who fixed it for twenty bucks," he marvels.

By noon the following afternoon, Jim has collected enough samples for the investigation. "I'll send these samples to the lab for analysis," he explains. "Based on the results, I should be able to determine whether the contamination is affecting the water supply and how we should go about cleaning it up."

"This was an unusual investigation," he adds. "Things don't usually go so smoothly." With a keen sense of satisfaction, Jim hops in his truck and heads back to St. Paul.

What Does a Groundwater Professional Do?

Unlike the water we see in streams, lakes, and oceans, groundwater is hidden beneath the earth. Before human beings can use groundwater for drinking water, they must find it, bring it to the surface of the earth, and remove any contaminants. This is exceptionally important work. As Jim Lundy explains, "Water is essential for life. To survive, we all have to drink a significant amount of water each day. About half of the water we drink in the United States today comes from groundwater."

No one in the working world actually bears the title "groundwater professional." This general term is used to refer to a variety of individuals who study the availability, quantity, and quality of groundwater supplies. Hydrogeologists, hydrologists, chemists, and engineers all can specialize in the study of groundwater. *Hydrogeologists* study the availability of groundwater supplies. *Hydrologists* study the composition, distribution, and movement of surface and underground water.

Lingo to Learn

Aquifer: *A water-bearing layer of permeable rock, soil, or unconsolidated glacial overburden.*

Bailer: *Used for collecting water samples from wells. Usually a plastic tube, open at either end, with check valves. The check valves seal as the bailer is pulled out of water, thereby collecting samples.*

DNAPL: *A dense non-aqueous phase liquid, or a liquid that tends to sink in water without readily dissolving. DNAPLs include solvents and some oils. Because DNAPLs sink in aquifers, they can be difficult to locate and remove.*

LNAPL: *A light non-aqueous phase liquid, or gasoline or oil that does not tend to sink in water.*

NAPL: *Non-aqueous phase liquid, or an oil.*

Plume: *An area of contamination within groundwater.*

Chemists figure out how different contaminants will behave in water and how these contaminants can be removed. *Engineers* design systems to purify groundwater and to distribute it to communities. Most groundwater professionals work in teams that include a variety of specialists.

In general terms, groundwater professionals locate water supplies, study water distribution, eliminate contamination, enforce environmental regulations, and help companies and communities comply with environmental regulations.

Groundwater professionals locate groundwater supplies by drilling into the earth and collecting samples. They do not, however, drill randomly. This would be an extremely expensive, time-consuming, and discouraging way to proceed. Instead, they study geological data, topographical maps, vegetation patterns, and a host of other natural clues to identify possible groundwater supplies. Once a groundwater professional has located an underground water supply, he or she must determine whether the supply is large enough to use for drinking water and analyze the quality of the supply. If the water is contaminated with harmful chemicals, industrial products, or saltwater, it may not be useful as a source of drinking water. The groundwater professional also must locate the source of the water in order to determine whether it can replenish itself quickly enough to be useful as a permanent source of drinking water. If a groundwater supply meets all of the necessary criteria, the groundwater professional may work with environmental engineers to design systems for moving and purifying the water.

Groundwater professionals who search for new water sources usually work for consulting companies that are hired by municipalities. Larger governing bodies, such as state or federal environmental agencies, may employ groundwater professionals to study and map the distribution of underground water supplies within a state or region or to predict the possible effects of contamination within a specified area. Groundwater professionals who map the distribution of water in a region must collect and study reams of data. Many use sophisticated computer programs to model the probable impact of rainfall, droughts, or contamination.

Some groundwater professionals specialize in removing contamination from groundwater supplies. Groundwater professionals today have a vast arsenal of weapons with which to combat contamination. They use microbes or bacteria to consume contamination. They run soil through thermal roasters to burn off contaminants. Another common method, called air stripping, consists of allowing volatile chemicals to escape into the air. These are just a few of

What Does a Groundwater Professional Do?, continued

the methods groundwater professionals can use to remove contaminants from groundwater or from surrounding soil.

Many groundwater professionals are concerned with protecting known groundwater supplies from contamination. Some of these individuals, most of whom work for government agencies, issue permits to developers. Before approving a developer's plans, a groundwater professional must consider the use that will be made of the land, its proximity to groundwater supplies, the likelihood that contamination will occur, and the probable impact of any contamination on surrounding communities. The groundwater professional also must monitor the builder's progress to ensure that proper measures are taken to avoid compromising nearby groundwater supplies.

A significant number of groundwater professionals are charged with enforcing environmental regulations. These individuals, usually hired by state or federal government agencies, require companies to submit regular reports demonstrating their compliance with regulations. If the groundwater professional questions a company's data, he or she can require additional testing, issue fines, or demand a change in procedures. Because companies are required to demonstrate compliance, many hire groundwater experts to monitor their procedures, to test soil and groundwater supplies adjacent to their operations, and to compile reports for regulatory agencies. Companies also turn to groundwater professionals for assistance in developing systems that minimize the possibility of groundwater contamination or in swiftly cleaning up any contamination that does occur.

What Is It Like to Be a Groundwater Professional?

Because this profession includes so many different types of specialists, one cannot easily generalize about the working conditions of groundwater professionals. It is safe to say, though, that most groundwater professionals combine indoor and outdoor work. Most also work in teams with groundwater professionals who have other areas of specialty. The vast majority of people in this field spend a significant percentage of their time summarizing results and writing reports.

Jim Lundy, a hydrogeologist employed by the Minnesota Pollution Control Agency, estimates that he spends 75 percent of his time in his office. As a government employee, Jim is responsible for enforcing state water regulations. Consequently, he spends a great deal of time evaluating the data and reports submitted by private industry, municipalities, and consultants. He also must write his own reports, assessing the data provided by these entities. "I also do some fieldwork," says Jim. "I sometimes supervise the drilling of monitoring wells. I decide where the wells should be drilled and how deep they should be, and I collect soil and water samples."

"I sometimes supervise the drilling of monitoring wells. I decide where to the wells should be drilled and how deep they should be, and I collect soil and water samples."

Jim's job also involves considerable research. "Right now, for example, I'm trying to locate the source of TCE—trichloroethene—in a town's water supply," says Jim. "TCE is a common industrial solvent used to clean parts. The level of TCE in this town's water is not dangerous but, because TCE is carcinogenic, we obviously are concerned about its presence in any community's water supply. We also know that TCE is not very stable underground—it degrades into other compounds that are even worse than TCE."

Jim will use a number of different strategies to track down the source of the TCE. He may start by visiting the library to study aerial photographs and newspaper clippings. He explains, "Aerial photos are a great resource. They can show how a property was developed over the years and where factories or storage tanks were located in the past. Sometimes you can spot a stain, or an area where vegetation does not seem to grow, both of which may indicate contamination. Newspaper clippings may mention a large fire or spill in the area, which may have been the cause of the problem."

Jim will also visit the site of the problem to look for clues. He will drive around the area, noting any commercial enterprises that may use TCE in their operations. "We always start with the cheap methods of investigation," he adds. "We try to narrow down the possible sources of contamination before we start drilling for samples."

Mark Hagley is a hydrogeologist who works for Barr Engineering Company, a Minnesota consulting firm. As a consultant, Mark provides services for both government entities and members of private industry. An experienced professional, Mark has significant managerial responsibilities. These

What Is It Like to Be a . . . ?, continued

days, in fact, Mark characterizes himself as "more of a project manager and less of a scientist."

"I schedule field events, manage staff, and oversee subcontractors for investigations and cleanup efforts," he says.

Mark's job offers tremendous variety. He estimates that he spends about 20 percent of his time on field activities, such as installing monitoring wells or wells that provide drinking water or conducting soil investigations. He devotes another 20 percent of his time to what he calls "real science," such as data evaluation and assessment. He spends 20 percent of his time writing reports. He spends another 20 percent of his time shepherding projects from start to finish. The remainder of his time is dedicated to administrative tasks, such as managing budgets, keeping records, and staying up to date on technological and regulatory issues.

Have I Got What It Takes to Be a GP?

"Groundwater professionals," says Jim Lundy, "must combine attention to detail with a strong ability to look at the big picture. They have to be able to consider possibilities. They should constantly ask themselves what might be happening."

"There's no Rosetta stone in this field," he adds. "Nothing is clear cut. Groundwater professionals must have a high tolerance for ambiguity and they must be able to develop a cogent argument from a few pieces of a puzzle. They also must be able to do several things at a time with a reasonable degree of accuracy."

Jim also believes that groundwater professionals must be decisive. He elaborates, "In this field, you've got to be able to make a decision. You have to know when you have enough information to make a decision and when to wait. You've also got to know when to admit that you've made the wrong decision."

Since groundwater professionals almost always work within teams, flexibility, cooperation, and excellent communication skills are also critical to success within this field.

To be a successful groundwater professional, you should:

Be attentive to the smallest of details without losing sight of the big picture

Like to solve problems

Be able to make quick, intelligent decisions

Be able to work as part of a team

Have strong writing skills in order to write reports, and excellent oral skills in order to work well with co-workers and others

How Do I Become a Groundwater Professional?

Groundwater professionals must have a bachelor's degree in a related area of study, such as geology, hydrology, civil engineering, or chemistry. A growing number of groundwater professionals today choose to pursue master's degrees in their areas of specialty.

Because our understanding of the behavior of chemicals within groundwater continues to grow, and our ability to remediate water continues to advance, groundwater professionals must plan to continue learning throughout their careers. Groundwater professionals must attend annual seminars and read publications such as the *Groundwater Journal* and *Groundwater Monitoring and Remediation* in order to stay abreast of new technical developments and environmental regulations.

Education

High School

High school students who are interested in preserving and restoring our nation's water supply should prepare themselves by studying science—particularly chemistry and geology—and math. They should also learn as much as possible about computers. Students also should take courses that will enhance their verbal and written communication skills.

> "I would be pretty impressed by a high school student who . . . knew enough to want to do an internship."

Motivated students also may be able to obtain relevant hands-on experience through unpaid internship positions. "Due to insurance restrictions, we can't send anyone who is not a full-time employee to a site," observes Jim Lundy, "but there is plenty of work that does not involve site visits. Students can conduct library research or look up old aerial photographs. There's always a way to set something like that up for a student who takes initiative."

He adds, "I would be pretty impressed by a high school student who thought to do something like that—who knew enough to want to do an internship."

How Do I Become a . . . ?, continued

Postsecondary Training

Once in college, students should concentrate on science courses, such as geology, chemistry, water chemistry, and physics. They also should study computer mapping and modeling. If possible, students also should study Geographic Information Systems, or GIS. Scientists use GIS to create spatial databases that enable them to analyze geographic locations from different views and cross sections. A background in foreign language will be helpful for future ground-

Silent Spring, *Catalyst for Change*

In 1962, Rachel Carson fired a literary warning flare into the sky. Her book, Silent Spring, *alerted people to the extreme danger of indiscriminate pesticide use. Carson argued that, if left unchecked, the dizzying pace with which science was introducing new chemicals, many of which were pesticides, would eventually make the Earth unfit for life. Her message stunned the nation.*

After World War II, when dichlorodiphenyltrichloroethane, or DDT, was used to prevent the transmittal of typhus through fleas, most people regarded chemical insecticides as miracle substances that could protect crops and prevent insect-borne disease. Because DDT and other chemical insecticides were inexpensive, long-lasting, and commonly believed to be safe, they were widely used to spray farms, public spaces, and even backyards.

In Silent Spring, *Rachel Carson explained that DDT and chemical pesticides remain in their toxic form for years, contaminating soil, streams, and groundwater. Carson stated that these substances were already killing birds and fish, upsetting ecosystems, and causing human nervous system disorders. Because these chemical compounds are stored in the fat cells of animals that ingest them, Carson believed that the widespread use of chemical pesticides would have long-term implications throughout the food chain. Eventually, she said, they would lead to devastating health problems in human populations.*

In response to Silent Spring, *state legislatures throughout the nation introduced bills to regulate pesticide use. The use of DDT was banned in the United States. In 1970, the United States Environmental Protection Agency (EPA) was created. The responsibility for pesticide regulation was moved from the Agriculture Department, which tended to see the advantages of these chemicals, to the EPA.*

The problem is far from solved, however. According to Vice President Al Gore, our nation used 2.2 billion pounds of pesticides in 1992, an amount roughly equivalent to eight pounds for each person living in our country. Many of these substances are highly carcinogenic. Others damage the nervous and immune systems of living organisms. These chemicals continue to contaminate our soil and the groundwater that becomes our drinking water.

Irresponsible pesticide use is a global problem. Nations around the world use pesticides, many with fewer restraints than those imposed in the United States. Many countries continue to use pesticides that have been banned in the United States, such as DDT. These chemicals are contaminating our planet's earth, water, and air.

Rachel Carson shattered our complacency about pesticide use. If we are to benefit from her warning, we must continue to fight for increased regulation of pesticides.

water professionals, as many companies expand their horizons to international remediation.

Internships and Volunteerships

Students also should seek field experience through internship opportunities. Since the number of students invariably exceeds the number of available internships, students should be aggressive and creative in their efforts to find opportunities. Students can begin by discussing internship possibilities with their professors. They should also contact consulting companies and government environmental agencies. Organizations often are willing to create internship positions for students who appear genuinely motivated.

Certification or Licensing

Certification for groundwater professionals is voluntary. A number of professional organizations offer certification opportunities, including the National Ground Water Association, the American Institute of Hydrology, and the American Water Resources Association. Students should contact associations that represent professionals in their area of specialty for additional information. Students also may wish to contact the environmental regulatory agencies in their states to determine which certifications are considered most useful. Many states also now offer certification.

Who Will Hire Me?

"Many of the groundwater opportunities today are in consulting firms," notes Mark Hagley. In fact, consulting companies account for about half of all groundwater-related positions.

Federal, state, and local government agencies employ a significant number of groundwater professionals to locate groundwater supplies, map groundwater distribution, monitor groundwater quality, and enforce environmental regulations.

Some very large companies employ full-time groundwater professionals, as well. Major petroleum companies, for example, may employ groundwater professionals to ensure that their underground storage tanks do not contaminate groundwater supplies. Most private companies, however, prefer to hire consultants to help them comply with environmental regulations. Consultants are usually considered less expensive than full-time employees because companies need not provide health insurance and other benefits for them and

Who Will Hire Me?, continued

because most companies only deal with groundwater issues on an irregular basis.

Colleges and universities also employ groundwater professionals to teach and conduct research. Most individuals who hope to pursue academic careers should obtain doctoral degrees, though a small number of groundwater professionals may find positions teaching general earth science courses at the secondary level.

Where Can I Go from Here?

Groundwater professionals can choose from many different avenues of advancement. Those who are interested in regular hours and comfortable salaries may assume administrative and managerial responsibilities within government agencies. Those who seek higher salaries, are willing to work long, sometimes irregular hours, and are not adverse to some financial risk, can assume partnership positions within consulting firms. Groundwater professionals who are interested in adventure, and in improving the lives of others, may use their skills to aid countries that are struggling to provide safe drinking water for exponentially growing populations.

As is true in most professions, groundwater professionals also can advance by continuing to enhance their skills and knowledge by developing an area of expertise or by obtaining certification.

What Are Some Related Jobs?

The U.S. Department of Labor classifies groundwater professionals with people who work in the physical sciences (GOE). This category includes geologists, hydrogeologists, hydrologists, petrologists, seismologists, and environmental analysts. The U.S. Department of Labor also classifies groundwater professionals with people who work in other geology occupations (DOT), such as geodesists, geophysical prospectors, mineralogists, paleontologists, and soil engineers.

Groundwater professionals are also closely related to, and usually work closely with, various engineering specialists, including civil engineers, environmental engineers, sanitary engineers, and chemical engineers.

WHAT ARE THE SALARY RANGES?

When asked how much money a groundwater professional typically earns, Jim Lundy deadpans, "Millions and millions."

With a laugh, he continues, "Actually, most groundwater professionals start out in the mid-twenties. After a few years of experience, they may earn between $30,000 and $40,000. Those at the top of the field, and in the private sector, can eventually earn up to $50,000 to $60,000 a year."

Jim's estimates are supported by figures compiled by the National Association of Colleges and Employers, which in 1997 found that a new employee with a bachelor's degree in geology or geological sciences earned an average salary of $31,000. A new employee with a degree in environmental science earned about $26,000.

Related Jobs

Chemical engineers
Civil engineers
Environmental analysts
Environmental engineers
Geodesists
Geologists
Geophysical prospectors
Hydrogeologists
Hydrologists
Mineralogists
Petrologists
Sanitary engineers
Seismologists
Soil engineers

"As long as people continue to need water, there will be a need for groundwater professionals."

According to a survey conducted by the Association of Ground Water Scientists and Engineers (AGWSE), the average salary in this field is $38,250. The Professionals in Science and Technology reports that hydrogeologists earn starting salaries of $34,942. Those with slightly more experience earn an average of $45,835 and those with advanced degrees and more experience earn an average of $58,878.

Generally speaking, consulting firms offer higher salaries than government agencies. Salaries also vary by region. Generally speaking, a groundwater professional employed on the east or west coast will earn more than a groundwater professional in Chicago or Minneapolis.

WHAT IS THE JOB OUTLOOK?

Like many environmental occupations, groundwater career opportunities surged in the 1980s in response to stricter government regulations. The field continued to grow through the mid-1990s but now is beginning to level off. Jim

What Is the Job Outlook?, continued

comments, "The field was burgeoning in the 1980s. I don't know whether we'll ever see growth like that again, but the field isn't going to contract severely either. As long as people continue to need water, there will be a need for groundwater professionals."

While this field may not be growing as rapidly today as it was five to ten years ago, it remains a promising career choice for motivated, intelligent students. The continued growth of our nation's population will make finding and remediating groundwater supplies an even more pressing issue in the next century, creating many interesting and rewarding opportunities for qualified groundwater professionals.

Hazardous Waste Management Technician

Definition
Hazardous waste management technicians *support the work of engineers and scientists in the identification, removal, and disposal of hazardous wastes. Technicians may collect water, soil, or other samples for laboratory analysis; monitor waste control systems; and perform other work, complying with federal, state, municipal, and other regulations.*

Alternative Job Titles
Emergency responder
Environmental technologist

Salary Range
$25,000 to $55,000 to $85,000

Educational Requirements
High school diploma
Job-specific training or two-year associate's degree

Certification or Licensing
Required

Employment Outlook
Faster than the average

High School Subjects
Biology
Chemistry
Computers
English (writing/literature)
Geology
Mathematics

Personal Interests
The Environment
Science

Late Friday morning, Sam Borries receives a call from the Indiana Department of Environment Management (IDEM). A petroleum leak has been discovered in Hammond, Indiana. Gas is escaping a petroleum pipeline and is seeping into the Hammond sewer system. IDEM personnel fear that, if not quickly contained, the leak could result in a fire or explosion or could contaminate the Little Calumet River. Sam, an on-scene coordinator for the United States Environmental Protection Agency (EPA), calmly questions the caller. He agrees to help coordinate the cleanup.

On the scene, Sam quickly assesses the situation. The leak had been discovered on Thursday, after members of the community reported smelling gasoline odors. The responsible petroleum company had moved quickly to identify the leaking pipeline and block it off. Though unsure how much gas had escaped into the sewer system, they immediately notified the appropriate authorities and began recovery efforts. They mobilized huge vacuum recovery trucks to suck the gas out of the sewer system. Because the gas vapors were extremely dense, they also closed the street above the affected lines. The cleanup is well underway, but the threat of a fire or an explosion persists.

With Sam's assistance, the petroleum company, IDEM, and the Hammond Department of Environmental Management (HDEM) mobilize crews to dig up the road in order to locate the leaky gas line. As they search for the source of the leak, crews use sandbags to try to contain the escaping gas. They use a product called Biosolve to suppress gasoline vapors. In the afternoon, the team attempts to open the road above the sewer to one lane of traffic in each direction. When a passing motorist tosses a lit cigarette from his window, however, the team quickly realizes that this is too dangerous and closes the road until the response effort is completed.

The cleanup effort continues throughout the weekend. On Monday, the process is complicated by reports of gasoline odors in homes along the sewer lines. The team realizes that the incident is larger than first believed. They hire a local marine diving company to find where the fuel is infiltrating the sewer line and to help devise a plan for plugging the leaks. By Wednesday, the petroleum company's pipeline has been repaired and is functioning normally. The emergency response team must still remove all the fuel and contaminated water from the sewer lines before repairing the road.

Lingo to Learn

Comprehensive Environmental Response, Compensation, and Liability Act (CERCLA): *1980 law (known as "Superfund") that mandated cleanup of private and government-owned hazardous waste sites.*

Defense waste: *Radioactive waste from weapons research and development, decommissioned nuclear-powered ships and submarines, weapons material production, and other military waste. Includes low- and high-level radioactive, hazardous, and mixed (radioactive and hazardous) waste.*

Level A work: *Work involving hazardous substances that pose a high dermal (skin) threat.*

National Priorities List: *U.S. EPA list of the worst hazardous waste sites in the country needing cleanup. Currently, there are 1,245 sites on the list, and cleanups of NPL sites are taking up to eight to ten years each.*

Sampling: *Taking quantities of water, soil, or air from a site. Samples are tested in labs to check for the presence of hazardous substances.*

U.S. Environmental Protection Agency (EPA): *Federal agency responsible for overseeing the implementation of environmental laws, including those designed to monitor and control air, water, and soil pollution. State EPAs help carry out these laws.*

WHAT DOES A HWMT DO?

Hazardous waste management technicians are trained to safely contain and remove highly toxic or volatile materials. Broadly defined, a hazardous waste is any substance that threatens human health or the environment.

Some hazardous waste management technicians respond to emergency situations, such as the petroleum leak in Hammond, Indiana. As an EPA employee, Sam's mandate is to respond to emergencies that pose "an immediate and substantial danger to human health and the environment." Sam, therefore, responds primarily to emergency situations, such as petroleum or chemical spills, though he also supervises ongoing cleanup efforts. Other hazardous waste management technicians clean up devastating industrial messes—ugly reminders of a time when our country did not monitor environmental contamination as closely as we do today. Such sites are specifically targeted for cleanup by the federal government under the Comprehensive Environmental Response, Compensation, and Liability Act (CERCLA) of 1980, known as "Superfund."

Since many of the worst Superfund sites have, today, been cleaned up, a larger number of hazardous waste management technicians are involved in lower priority cleanups. Many are cleaning up previously abandoned industrial sites, called brownfields, so that the land can once again be used. These sites do not pose an immediate threat to human health or the environment. "As the country learns more about the effects of hazardous waste on the environment," notes Carl Patterson, a technician with the consulting firm CH2M Hill, "we're going in and cleaning up those kinds of sites, too."

The Environmental Careers Organization (ECO), a major environmental careers internship and job placement specialist in the United States, estimates that there are up to 33,000 municipal and industrial hazardous waste sites in the United States, plus 5,000 to 10,000 government-owned sites. In addition to helping clean up these kinds of sites, hazardous waste management technicians may identify, categorize, and dispose of hazardous waste, or help in efforts to lower the amount of hazardous waste produced in the first place. They also assist in other types of cleanup and pollution prevention efforts, including routine monitoring of air, soil, and water, to make sure hazardous substances are within acceptable limits.

Nearly every industry produces hazardous waste—from food, textiles, metals, petroleum, plastics, and paper manufacturing to dry cleaning services, printing, and more. Chemical companies produce up to 68 percent of all private-industry hazardous waste. Petroleum companies are another significant source of hazardous waste. For example, one common hazardous waste is ben-

WHAT DOES A HWMT DO?, CONTINUED

zene, a component in fuel that has been identified as a carcinogen. Tanks of fuel around a gas station can leak, and substances from the fuel, including benzene, can seep into the ground and contaminate the groundwater—which we then drink. Other hazardous wastes include solvents used in paint, manufacturing, and service operations such as dry cleaners. Manufacturing processes also produce certain metals that are hazardous.

The federal government is also a leading generator of hazardous waste. According to the ECO, domestic military bases produce more hazardous waste than the top five chemical companies combined. The U.S. Department of Defense (DOD) spends about $5 billion per year on hazardous waste cleanups. Estimates vary, but all together, between 5,000 and 10,000 government-owned sites require cleanups—at a projected cost of $250 billion. The government usually hires private environmental consultants to carry out its cleanup jobs, which include military bases and military production sites.

The United States has some of the toughest environmental laws in the world. Beginning in the 1960s, people began to realize that this country's industry, including its military production industry, was producing vast amounts of pollution and waste that was ruining the environment and threatening human health and safety. Private industry, municipalities, and the government had been producing hazardous and other wastes for many years. A series of tough laws, including CERCLA, have been passed over the last twenty-five years to force the cleanup of old waste sites and discourage companies from creating new ones. Although the cost of cleaning up our environment is high—billions of dollars have been spent to clean up hazardous waste sites—the cost of not cleaning up is potentially even higher.

WHAT IS IT LIKE TO BE A HWMT?

In his role as an on-site coordinator for the EPA, Sam Borries divides his time between his office and the sites of major spills. Sam's job is to coordinate the federal response to hazards. He typically coordinates his efforts with state and local authorities, including police and fire departments. If the responsible authority does not have the ability to adequately respond to a situation, however, Sam can step in and assume complete responsibility for the project. If local authorities have a situation under control, on the other hand, Sam may oversee the progress from his office.

"I respond to a lot of petroleum leaks," says Sam. "Under the Oil Pollution Act, I get involved when oil is released to a waterway. I may be working on a river, lake, wetland, or a storm sewer. It just depends on where the spill is."

When an emergency occurs, Sam's first priority is to put out any fires that occur. Once the fires have been extinguished, he tries to identify the source of the leak or spill in order to stop the release. The next step is to control and contain the hazardous material that has already been released. Once the substance has been contained, Sam and his colleagues devise a strategy for cleaning it up. "There are a variety of tools for cleaning up petroleum products," Sam notes. "We can use skimmers or vacuums to wick the petroleum off the water. We can run the water through an oil-water separator. Usually we try to treat the oil so that it can be reused. Sometimes we have to burn it off the water, though this is a less desirable method."

"People want to know if there's a health threat. Their first question is whether they've been exposed to anything and, if so, what they can do. Then they want to know how the situation will disrupt daily life. I've been in front of TV cameras more times than I can tell you."

In addition to his technical responsibilities, Sam must perform public relations duties throughout emergency situations. "People want to know if there's a health threat," he explains. "Their first question is whether they've been exposed to anything and, if so, what they can do. Then they want to know how the situation will disrupt daily life. I've been in front of TV cameras more times than I can tell you."

Carl Patterson is a technician for CH2M Hill, one of the largest environmental consulting firms in the United States. CH2M Hill does every type of environmental analysis, cleanup, and follow-up project for private industrial companies, municipalities, and the government. CH2M Hill is also part of a global organization that does water plant and environmental projects all over the world.

Carl is based in the company's Deerfield Beach, Florida, offices. He helps to do the highly accurate measuring, monitoring, and sampling needed

WHAT IS IT LIKE TO BE A . . . ?, CONTINUED

to be sure that a client is complying with environmental laws. Now in his sixth year with CH2M Hill, Carl focuses on projects to identify, correct, prevent, and monitor the contamination of water by hazardous substances. He is concerned with both surface water, such as lakes, streams, and canals, and with groundwater. When a fuel or chemical spills onto the ground, hazardous substances can soak through the soil and reach the groundwater, contaminating it. This is extremely dangerous, because groundwater is used for community water supplies. "Water treatment facilities aren't designed to handle these kinds of hazardous substances," Carl says. "And in some places, like outlying rural areas, people are drinking groundwater, drawing it right out of the ground from wells. The efforts of federal, state, and local authorities to provide safe drinking water are really incredible—especially in this area. Florida's environment is very wet and marshy. The groundwater is very close to the surface—usually just a few feet. There's a high potential for problems. Also, there are miles of canals to control flooding, and that means a lot of exposed groundwater with potential for contamination."

"For example," Carl continues, "there's a fuel storage facility near here. They have thousands of barrels of fuel. Obviously, there's a huge potential for a problem. The state requires the facility to monitor the groundwater around that site, so a daily task for us is to take groundwater samples from the monitoring wells there. CH2M Hill has its own labs; they run tests on the samples and see if there are hazardous substances in them. If so, and if there's more of the substance than allowed by federal, state, or local authorities, CH2M Hill can tell owners immediately. We also can help them develop a plan to clean up the problem and take steps to prevent other leaks."

To collect water samples, Carl uses a narrow, three-foot-long, tube-like instrument made of Teflon or stainless steel. Water is drawn in at the bottom of the instrument, following strict procedures defined by the EPA, the county in which the site is located, CH2M Hill, and sometimes the client. CH2M Hill's job is to be so precise with the retrieval procedures and in the analysis done back at the labs that the company produces "defendable data," such as data that can stand up in court. "I've never seen anything go to litigation," Carl says, "but that's how we operate—in terms of potentially defending in court the processes we used."

Accuracy is extremely important. "We work with very tiny numbers, very tiny compounds," he says. "For example, for benzene, one part per billion in the water is considered too much. When the test results are close, we'll go back and re-test—particularly when there's a potential for very costly remedial projects. In fact, almost everything is triple redundancy [done three times]."

Altogether, Carl runs about one hundred field tests a year. "Essentially, we do fieldwork for one to two months out of the year. The rest of the year is spent compiling data." Carl's work takes him to a wide variety of sites. "It's hard to express how broad it is, how there are projects almost everywhere," he says. "Sites are indoors and outdoors, and range from the obvious, like an oil refinery, to the less obvious, like a carpet factory. Metal manufacturing can create dust from cutting, bending or other processes; when you sweep or wash the dust out of the plant, it has the potential to reach the groundwater and contaminate it."

"Some really smell bad. Some of these places are producing some really nasty effluents. I'm sometimes stunned at what people do with their wastes."

Technicians who work in the field, testing and monitoring air, soil, or water, often enjoy being outdoors, free of the routine of office or lab work. But not all sites are outdoors, and many sites are unpleasant. "Some indoor sites are so noisy that you can't communicate. By the end of the day, you're frazzled," Carl says. "Some really smell bad. Some of these places are producing some really nasty effluents. I'm sometimes stunned at what people do with their wastes.

"Sometimes you're all bundled up, in special suits, a paper hat, a respirator," he continues. "The respirator's just an air-purifying mask with filters, but I'm claustrophobic and it bothers me. It makes it difficult to communicate. The ones I can't deal with are the zip-on suits. You're sealed in a suit from head to toe, like an astronaut. They're for Level A work, for substances with a very high dermal threat—those that will burn your skin." Luckily, he adds, CH2M Hill doesn't do Level A testing.

"There's lots of travel to this job, and sometimes real travel, all over the world," he says. "But sometimes you're traveling only to see the armpit of a country—the nasty side. And the workday is long."

WHAT IS IT LIKE TO BE A . . . ?, CONTINUED

Still, there is a lot about the job that Carl really likes. "Each project is almost a new world unto itself," he says. "And I like being part of the cleanup efforts—after all, it's my drinking water, too."

HAVE I GOT WHAT IT TAKES TO BE A HWMT?

Hazardous waste management technicians work with toxic and volatile materials. To ensure their own safety, and that of others, they must be extremely alert and accurate. They also must be able to follow orders. Some, but not all, positions require technical and scientific aptitude. Hazardous waste management technicians usually work in teams, so they must be able to cooperate and communicate well with others. Good reasoning skills, analytical thinking, and the ability to respond to unexpected conditions are also important.

To be a successful hazardous waste management technician, you should:

Be willing to work with toxic and sometimes volatile materials

Have technical and scientific aptitude

Be able to work as a member of a team

Be a good listener and know how to follow rules—standards and regulations set by your employer and all levels of government

Be flexible and be willing to be on call and go to work at a moment's notice

Be able to think quickly on your feet

In many ways, technicians must be rule-followers, and they must be careful listeners. Regulations drive this industry—there are standards and guidelines that must be followed for every part of the hazardous waste management technician's job. If a standard requires the hazardous waste management technician to collect a half-liter sample, he or she must get exactly that. When hazardous waste management technicians do not follow regulations, the price can be high. They can cost their employers or clients a great deal of money in fines, or, worse, they can endanger their lives and the lives of others.

"People in this profession must be motivated and they must have excellent people skills," says Sam Borries. "They also must be disciplined. They have to be willing to go to work at a moment's notice. Like most hazardous waste management technicians, I take turns being on call. Someone has to be on call 365 days a year. When I'm on call, I may get called in the middle of the night and, if I do, I go."

Good technicians also must be flexible, able to spot problems, and able to think quickly on their feet. "Sometimes, once you get out to the site, you find that the ground's not dirt like everyone expected—it's mud," Carl Patterson says. "There were plans for specific tests, but now it's pouring rain. Or there are power lines that the managers or engineers didn't anticipate. You have to be flexible and able to adapt to the conditions you find."

Good technicians are conscientious. "This job is all about details," Carl says. "But I also like to think about the big picture. For example, sometimes at a cleaned-up site there'll be ground cleaners and incinerators. It's a standard setup: they'll have a system to drag water out of the ground, run it through stripping towers, and then pump it back in the ground. It helps to be able to picture how that's working. I need to know if a system is there, if it's on or off, and if it will change the way the tests should be done."

How Do I Become a HWMT?

In the past, hazardous waste management technicians could find plenty of work with only a high school diploma. Hazardous waste management is becoming an increasingly sophisticated field, however, because of tighter regulations and advances in cleanup technology. More and more, a two-year diploma or degree in hazardous waste management is becoming important for many positions.

The Exxon Valdez Oil Spill

Just past midnight on March 24, 1989, an enormous oil tanker, the Exxon Valdez, ran aground on Bligh Reef in the northeastern part of beautiful Prince William Sound. Over the next few days, the tanker spilled more than eleven million gallons of crude oil into the surrounding waters, creating an oil slick that measured seven miles long and seven miles wide. It was the largest oil spill in U.S. history.

More than 11,000 people, many of them hazardous waste management technicians, contributed to the cleanup effort. In the weeks and months that followed the spill, people worked around the clock to try to contain and remove the oil. They used an array of cleanup techniques, including burning, chemical dispersants, high-pressure/hot-water washing, cold-water washing, fertilizer-enhanced bioremediation, and manual and mechanical removal of oil and oil-laden sediments.

Despite the heroic efforts of these dedicated individuals, the formerly pristine waters of Prince William Sound were severely contaminated. Between 100,000 and 300,000 birds died. Nearly 3,000 sea otters were killed. Alaska's fisheries, a major source of income for local communities, were severely impaired. Today, the marine life and other wildlife of Prince William Sound are still recovering.

The Exxon Valdez tragedy illustrates a harsh reality of environmental hazards. No matter how sophisticated the techniques for remediation become, hazardous waste management technicians can never totally eradicate the damage caused by environmental contamination.

How Do I Become a . . . ?, continued

"Whether or not a degree is required depends on the task, the company, and the nature of the problem," says Steve Wilson of the Air and Waste Management Association (AWMA). "For some field or monitoring work, two-year degrees may be needed. Some of these jobs involve sophisticated work like chemical analyses or working under protocols."

"On the other hand, other technicians are essentially moving waste, like forklift drivers, warehouse workers, or drivers," he adds. "They will get some particular training or instruction from the company, but they don't need a degree for this kind of work."

Experts see a trend toward higher educational requirements for environmental technicians overall than in the past. According to the Environmental Careers Organization, this is especially true in hazardous waste management. In this area, technical degrees—even graduate degrees—tend to be valued more than in some other areas, such as solid waste handling.

Education

High School

While still in high school, potential hazardous waste management technicians should take math and science, including biology, earth sciences, physics, or chemistry. To hone communication skills, they also should take English, speech, and writing.

"Grounding in the basic disciplines is going to be crucial," says Steve Wilson. "You may not be able to take all of them, but biology, chemistry, physics, the natural sciences, algebra, trigonometry, and geometry are very important."

Postsecondary Training

The United States Environmental Protection Agency (EPA) provides grants for schools to develop and implement environmental training curricula. Today, there are hundreds of choices for those interested in pursuing postsecondary training in hazardous waste management. Options include community colleges, technical colleges, vocational institutes, and college outreach programs. Students should make sure the school is accredited, and talk to the people in the placement office to find out where graduates have gone on to work.

Internships and Volunteerships

Students also should seek opportunities to gain firsthand experience by volunteering or participating in internships. Two national nonprofit groups help develop and place people in internships in the environmental industry: the

Student Conservation Association (SCA) and the Environmental Careers Organization (ECO). The SCA focuses on internships in resource management, placing people in projects in federal, state, and private parks and natural lands. There is a three- to five-week summer internship program for 450 high school students each year.

The ECO encourages everyone to get experience while in school through internships, volunteering for local nonprofit or community organizations, and independent research and study projects. The ECO offers internship opportunities for those at the college level or above. Created in 1972 with a handful of interns, ECO today is a major environmental placement service that also holds career seminars, recruits on campus, and recently began staging an annual Environmental Workforce Symposium. College students, recent grads, and first-time job seekers are matched with paid internships in the environmental industry. Many of the interns are eventually hired full-time and positions do include work in hazardous waste management. ECO publishes a book that outlines work throughout the environmental industry, *The New Complete Guide to Environmental Careers.* Both of these associations have Web sites. Check our "Surf the Web" section for further information.

Another option is to contact a federal or local government agency directly about an internship. Many of the federal-level agencies have internship programs, including the EPA, National Park Service, and Bureau of Land Management. Programs are more informal at the local level; contact regional, county, or other local authorities and express your interest in an internship. In the private sector, an internship in a nonprofit organization is probably the easiest to come by. These groups include the National Wildlife Federation and the Natural Resources Defense Council. Opportunities exist at both the local and national levels.

CERTIFICATION OR LICENSING

The Institute for Hazardous Materials Management gives a test for certifying people as hazardous materials managers. The National Environmental Health Association certifies hazardous waste specialists. In addition, a variety of accrediting bodies award professional engineering technician credentials.

WHO WILL HIRE ME?

Hazardous waste management technicians are employed by chemical companies and other producers of hazardous waste; waste disposal companies and

WHO WILL HIRE ME?, CONTINUED

waste disposal consulting engineering firms; environmental consulting firms; government agencies; and other organizations. By far, the largest number of jobs is in the private sector. Private companies employ about 75 percent of all employees in hazardous waste management, and public organizations (federal, state, and local) employ 24 percent, according to ECO. Nonprofit organizations employ the remaining 1 percent.

Following are some private-sector employers.

In-house staffs. Private industry jobs can be found within large companies. Such companies generate waste and are likely to have their own in-house staff of environmentalists. This is especially true as regulations keep getting more and more complex. Medium-size companies may have smaller departments. Smaller companies may have a professional or two on staff, or hire outside consultants.

Consultants. Consulting companies are another good source of employment opportunities for hazardous waste management technicians. Some consulting companies advise companies how to handle a hazardous waste problem. Others also design a plan and provide the manpower to carry it out. Some have their own testing and laboratory services. There are about one hundred very large environmental consulting firms in the United States, including companies like CH2M Hill, plus hundreds of smaller operations, including one-person outfits.

Disposal companies. Regulations for hazardous waste usually include directions for transporting, treating, or disposing of the waste. So special companies have evolved that do nothing but dispose of other companies' waste for them.

Following are public-sector employers.

The EPA is just one federal government agency that uses technicians. In fact, says ECO, the U.S. Forest Service, the National Park Service, the Bureau of Land Management, and the Fish and Wildlife Service probably employ more technicians and field personnel, while the EPA uses more scientists and other professionals. There also is a trend now toward increased work in hazardous waste management at the local level—by states and counties or municipalities. Jobs here include technicians at municipal water plants and other public facilities.

According to the ECO, a growing part of the hazardous waste management field is the handling and disposal of medical wastes. Hospitals, labs, health care facilities, and pharmaceutical companies may have staff personnel to help them take care of their medical wastes. Or, they may hire consultants to do the job. Smaller generators of hazardous wastes include university research facilities and even households. Another source of hazardous waste is inactive mines: hazardous minerals can leak into nearby surface and groundwater, creating potential health hazards.

WHERE CAN I GO FROM HERE?

Technicians typically don't go on to earn professional degrees, such as engineering or chemical degrees. However, there are several other advancement options that may be of interest to technicians. They may, for example, opt to specialize in the disposal of hazardous waste. These people conduct studies on hazardous waste management projects and provide information on treatment and containment of hazardous waste. At the government level, they help to develop hazardous waste rules and regulations.

"There also are incident commanders," says Mike Waxman. "An incident commander is the person who's in charge of and has ultimate responsibility for a hazardous waste site. They delegate tasks and interact with state and federal regulatory authorities as necessary. Basically, they carry the responsibility for the success or failure of the activity at the hazardous waste site."

Obtaining more education and training can help the technician earn more money and take on more responsibility. In some companies, additional education also will earn the technician a higher title. This is true at CH2M Hill, for example. "If I had a two-year associate's degree in one of the related scientific areas, my title would be different," Carl Patterson says. "If I had one in geology, for example, I'd be a geo-tech."

Some environmental professionals work in community relations or public affairs, helping to inform the public about what a company is doing with its wastes. The government also employs such professionals to help spread information about regulations and cleanup efforts.

WHAT ARE SOME RELATED JOBS?

The U.S. Department of Labor classifies hazardous waste management technicians under the heading, Inspectors and Investigators, Managerial and Public Service (DOT). Jobs closely related to hazardous waste management include solid waste management technicians, industrial safety and health technicians, wastewater treatment plant operators, land and water conservation technicians, landfill operators, physical geographers, pollution control technicians,

WHAT ARE SOME RELATED JOBS?, CONTINUED

radiation-protection specialists, safety inspectors, gas inspectors, sanitation inspectors, radioactive waste disposal dispatchers, and waste disposal attendants.

Other related occupations include *municipal water treatment plant technicians,* who test and monitor water quality at treatment plants to be sure it's safe to drink, and *radioactive disposal attendants,* who collect radioactive waste and contaminated equipment, load it on a truck using a forklift, and transport it to a storage facility or landfill. These workers may help make the concrete forms used to contain waste that will be buried. They also may clean contaminated equipment for re-use, using sand blasters and scrubbers, detergents, and other cleaning agents.

Related Jobs

Gas inspectors
Industrial safety and health technicians
Land and water conservation technicians
Landfill operators
Municipal water treatment plant technicians
Physical geographers
Pollution control technicians
Radiation-protection specialists
Radioactive disposal attendants
Safety inspectors
Sanitation inspectors
Solid waste management technicians
Wastewater treatment plant operators

WHAT ARE THE SALARY RANGES?

There are not yet many professional associations or national unions for hazardous waste management technicians, so there are few documented salary surveys. However, people in the field say that salaries range from good to excellent.

"Salaries vary a lot, depending on the region, the company, the specific duties of the technician, and his or her education and experience," says Mike Waxman. "Someone with no experience but with education beyond high school might start at about $25,000 to $30,000 a year. Someone with five years' experience plus a two-year degree in the field might make $30,000 to $50,000 a year."

ECO backs these figures, reporting starting salaries for hazardous waste management employees of $23,000 to $35,000 per year, ranging from $40,000 to $50,000 for experienced personnel, and $85,000 per year or more for top management positions. According to the ECO, consulting companies and in-house departments generally pay better than government agencies.

According to Sam Borries, hazardous waste management technicians who have a bachelor's degree and are employed by the EPA can earn from

$25,000 to $65,000. Managers with advanced training and education earn salaries that range from $65,000 to $85,000.

"Salaries do tend to vary by geographical region," Sam notes. For example, according to an August 1997 survey in *Environmental PROTECTION,* hazardous waste management technicians in the South average the highest salaries at $56,352 per year, while technicians employed in the West average the lowest at $54,185 per year.

WHAT IS THE JOB OUTLOOK?

"This field grew like gangbusters in the 1980s," says Sam Borries. "But it's leveling out now. I think the industry as a whole has matured."

The environmental industry is growing more slowly now than in the past—from 16 to 30 percent per year in the mid- to late 1980s to about 2 percent in 1991, 4 percent in 1992, and 5 to 7 percent in the late 1990s. *Environmental Business Journal* reports that hazardous waste management is growing faster than the overall industry; however, at rates of up to 16 to 18 percent through 1996. Continued growth is expected and will depend on political support for regulations, changes to regulations, the strength of the economy, and other factors.

What Are the Risks?

Are there significant health risks for people who work with hazardous waste? "When you put it that way—significant health risks—no," Carl says. "Because I have the training, equipment, and unbelievably redundant systems to protect me. My company has a comprehensive quality control plan that covers general decontamination procedures and so forth. Then, there are procedures for specific sites, such as a gas station site or a metals manufacturing plant."

"And in the field, the company gives me the right to shut down the project if I ever feel there's a threat to my safety or others'," he adds. "A lot of time and money go into planning a project; I better have a good reason for shutting it down. But if we go out in the field and see a problem that would cause some risk to our health, we go home. Even if it's a multimillion dollar project."

"In this job, there's (1) safety and (2) accuracy. Both are important. But if you have to put one first, the company says put safety first." "Sometimes you don't always know what's out there. Some chemicals have no smell, no taste. You can be breathing something [harmful] and you don't even know it. In my opinion PCBs are some of the worst. They're big molecules, but they like to grab onto things. You have to be careful. When you're working with water, something's always going to spill; with PCBs, if you're talking with someone, not paying attention, and the water splatters and gets on you're lip, you could inhale the PCBs. We decommission everything [equipment, clothing, vehicles] so we don't bring hazardous substances home with us."

WHAT IS THE JOB OUTLOOK?, CONTINUED

While they can expect good job demand and stability, technicians also should expect changes during their career because of advances in technology. "For example, the industry is looking at some things that would eliminate technicians going out as a sampler," says Carl Patterson, "like fiber-optic detectors that tell you directly what constituents are in the groundwater, so you don't need to do it in the lab. It saves money—there's no shipping, no transporting, no lab."

"But I think technicians will always be able to move into another area," he adds. "And some testing will always require a human presence. For example, there are mobile rigs to dig fifty to one hundred feet into the ground to take a sample. Someone needs to be there to be sure the rig is set up properly, with a tight seal, and be around when it's time to pull up the sample."

National Park Service Employee

Definition
National Park Service employees *work within the National Park Service in jobs dedicated to preserving our nation's natural and cultural resources and sharing them with the public.*

Alternative Job Titles
Backcountry ranger
Concessions specialist
Interpreter
Park planner
Public affairs officer
Resource manager

Salary Range
$20,233 to $25,897 to $70,000

Educational Requirements
High school diploma
Bachelor's degree

Certification or Licensing
None

Employment Outlook
Little change or more slowly than the average

High School Subjects
Anthropology and Archaeology
English (writing/literature)
History
Physical education
Sociology
Speech

Personal Interests
Animals
Botany
Business Management
Camping/Hiking
The Environment
Science
Teaching
Wildlife

The sun sits low in the sky, illuminating the red and gold that streaks the fantastic rock formations in Arches National Park in Utah.

Jim Webster pauses to search the horizon, squinting in the late afternoon light. He and the other rangers at Arches have joined the local police in searching for a woman who has been missing for four days. It is late December and desert nights can be cruel in winter. The search is becoming desperate.

The woman, a resident of nearby Moab, drove to Grand Junction to do some Christmas shopping. When she did not return, the local authorities began an exhaustive search. Helicopters fly over the land between Moab and Grand Junction. Members of the police department search the roads she may have taken. The rangers walk the parameters of the park, hoping to find some sign of the missing woman.

Late in the day, the police locate her car, stuck in a ravine just five miles north of the park. At last, they are able to focus the search. Jim and a team of search dogs begin combing the northern edge of the park. Before dusk, Jim spots a faint trail of footprints on the hard desert terrain. Sending the dogs racing forward, Jim radios the police. "We may have found her," he announces.

When at last Jim reaches the woman, she is dehydrated and near exhaustion, but she is alive. A helicopter arrives to transport her to a nearby hospital, and Jim resumes his ordinary responsibilities as the chief ranger of Arches National Park.

WHAT DOES A NATIONAL PARK SERVICE EMPLOYEE DO?

Our country's National Park System spans the country, covering more than eighty million acres. With only one exception (Delaware), every state in our country is home to at least one national park. Most of these parks welcome hundreds of thousands of visitors each year. To keep this amazing organization running, the National Park Service employs more than nine thousand permanent employees. An additional eleven thousand seasonal employees help out during the peak visitation season.

National Park Service employees have a wide variety of backgrounds and capabilities. They include law enforcement rangers, interpreters, resource managers, clerical assistants, maintenance workers, scientists, archaeologists, and historians—to name just a few. No matter what their responsibilities, these employees all are dedicated to achieving the National Park Service's mission, which is to conserve the natural and cultural resources of the National Park System for the enjoyment, education, and inspiration of this and future generations.

Each of these individuals performs an essential function within the park system. *Maintenance workers,* for instance, remove litter and keep the parks clean and beautiful. They also groom hiking trails, repair potholes, and restore historical buildings. Were it not for these dedicated, hardworking individuals, our parks would soon deteriorate. Our nation's pre-

Lingo to Learn

Alien species: *Species of plants or animals that do not naturally occur in a given area but were introduced by humans.*

Cultural resources: *Human artifacts or structures that are of historical or archeological significance.*

Endemism: *The existence of certain plant and animal species in only one location, usually in geographically isolated areas.*

Natural resources: *The wildlife, vegetation, or geological features within a national park.*

Speciation: *The natural development of new species.*

cious natural resources would be trampled and millions of park visitors each year would be disappointed.

Scientists, historians, and *archaeologists* are behind-the-scenes heroes within the National Park system. Scientists help us better understand the ecosystems within our parks, so that we can manage and use them more wisely. By studying the cultural artifacts within our parks, historians and archaeologists are able to help visitors learn about our country's past, the momentous events that shaped our nation, and the way our natural resources influenced those events.

The National Park Service employees who have the most contact with visitors are the *park rangers.* Though all rangers are trained to respond to emergency situations, there are actually two distinct kinds of rangers: those who enforce the rules and protect the park resources and those who interpret the resources to the public.

Enforcement rangers patrol the vast expanses of our nation's national parks, helping visitors have safe, enjoyable experiences in the wilderness. They are responsible for visitor protection, resource protection, law enforcement, and overseeing special park uses, such as commercial filming. They also collect park fees, provide emergency medical services, fight fires, and conduct wilderness rescues. In order to perform their responsibilities, they must spend a great deal of time in the field. Fieldwork may involve hiking the park's trails, patrolling the park's waters in boats, or interacting with visitors.

"Our job is to interpret the resources for visitors. We educate people about the value of our resources so that they appreciate them and want to take care of them."

Interpretive rangers are responsible for helping visitors understand the cultural and natural resources within our national parks. They try to educate the public about the history and value of the resources. They also try to help visitors learn how to have enriching, enjoyable experiences in the parks without harming the resources. Interpretive rangers give presentations, lead guided hikes, and answer questions. Some conduct orientation sessions for visitors as they first enter the park. Some also give presentations before community

WHAT DOES A NPS EMPLOYEE DO?, CONTINUED

groups and schools in order to help neighboring communities appreciate their parks. "Our job," explains Carol Spears, chief interpretive ranger at Channel Islands National Park in California, "is to interpret the resources for visitors. We educate people about the value of our resources so that they appreciate them and want to take care of them."

The many employees and functions within each national park all are overseen by one individual. This person, called the *park superintendent,* is charged with making sure that our parks maintain the delicate balance between welcoming visitors and preserving natural resources. He or she may, in larger parks, work with an *assistant superintendent.* In addition to supervising the various operations within a park, the superintendent handles land acquisitions, works with *resource managers* and *park planners* to direct development, and deals with local or national issues that may affect the future of the park.

WHAT IS IT LIKE TO BE A NPS EMPLOYEE?

As the chief ranger of Arches National Park, Jim Webster doesn't get to spend as much time in the field as he might like. Instead, he spends the majority of his day behind a desk, examining budgets, revenue figures, and personnel schedules. As manager of a department of 8 rangers, Jim must make sure his rangers are adequately patrolling the 73,379 acres of Arches. He hires, trains, and supervises the rangers that help visitors safely explore the exotic, arch-shaped rock formations that give the park its name.

Prior to becoming the chief ranger at Arches, Jim was an enforcement ranger. He has worked in a number of parks, including Yosemite, Grand Canyon, and Everglades and is very familiar with the life of a field ranger. "The parks must have some rangers available to visitors at all times, so most rangers either live within the park, or spend four days in the park followed by three days out of it," he says.

Enforcement rangers also must arrest drunk drivers or people who are stealing or damaging a park's resources, and ticket people who are violating traffic laws. In many parks, rangers also must fight fires. Jim recalls fighting fires when he worked at Crater Lake National Park in southwestern Oregon. "For big fires we would use shovels, hoses, and helicopters that sprayed water. Fire fighting can be extremely dramatic, but sometimes, when you're just shoveling dirt for hours on end, it can be really boring."

Like Jim, Carol Spears, chief interpretive ranger at Channel Islands National Park, spends most of her day behind a desk, on the phone, or in meetings. In addition to supervising the interpretive rangers, Carol contributes to long-term planning for the park. Although she misses her days in the field, Carol enjoys the planning aspects of her job. "I am able to plan for the park's long-term well-being," she explains. "I have to consider what is best for both the visitors and the resources. I have the knowledge that what I am doing is setting a course that will be followed for years to come. It gives me a real sense of purpose."

"I have the knowledge that what I am doing is setting a course that will be followed for years to come. It gives me a real sense of purpose."

To illustrate her point, Carol recalls the planning involved in a recent land acquisition. "We acquired 57,000 acres on Santa Rosa Island, which is 30 miles off shore. We wanted to open parts of the beach to camping, but we realized that the resources on Santa Rosa, as on most islands, are very fragile. The beaches of Santa Rosa are used by shorebirds, including a threatened subspecies of the Snowy Plover. We wanted the public to be able to enjoy the beaches, but we also had to protect the resources."

After countless meetings and extensive research, Carol and her colleagues arrived at a compromise. The beaches are closed during certain times of the year to allow the shorebirds to nest. During the rest of the year, campers, in limited numbers, are allowed on the beaches.

Prior to advancing to the chief interpreter position, Carol served as an interpretive ranger in a number of different parks. "I used to lead guided hikes to show people deer or other animals. At night, I sometimes led 'owl prowls,' on which I led visitors through the woods to watch the owls diving and calling out," she recalls. "I loved getting people excited about discovering the parks. It's very rewarding to see the excitement in the faces of visitors when you are introducing them to resources."

The interpretive rangers at Channel Islands face unusual challenges. "Channel Islands is very unique because one half of our resources are under water," says Carol. "Our rangers must patrol the water in boats. They talk to

WHAT IS IT LIKE TO BE A . . . ?, CONTINUED

boaters and divers to make sure that no one is damaging or taking marine animals. They also lead guided dives and conduct live underwater video programs for the public.

"It's very difficult to speak clearly and audibly when you're breathing through a mask," she adds with a laugh.

In addition to her love for the resources, Carol values the camaraderie among park rangers. "There are only four thousand permanent rangers within the Park Service," she explains. "So we have a sense of belonging to a family. There really is a kind of *esprit de corps* among National Park Rangers."

HAVE I GOT WHAT IT TAKES TO BE A NPSE?

National Park Service employees must successfully combine two very different characteristics. They must have a keen appreciation for nature and they must enjoy working with the public. Carol Spears explains, "As national park employees, we really have two missions. We must preserve the resources and we must provide for visitor use. Many times these two missions are in conflict with one another. We have to find ways to make them both happen."

Because most national park employees deal extensively with the public, they must be friendly, confident, and able to communicate clearly. Since they usually are responsible for a wide variety of tasks, they also must be exceptionally versatile. The fact that they work closely with nature, which can be unpredictable, means that these people must be creative problem solvers.

In addition to these general requirements, each of the positions within the National Park Service also involves a set of characteristics and abilities that are unique to that position. Superintendents, for instance, must be excellent administrators and must have the vision to make long-term plans for a park. Rangers must be able to react quickly and effectively in crisis situations and they must be able to convey authority to individuals who are violating park rules. Interpreters must have extensive knowledge about the resources in their parks and must be excellent educators.

To be a successful NPS employee, you should:

Love nature and enjoy working with the public

Be friendly, confident, and able to communicate clearly with co-workers and the general public

Be willing to work in a variety of natural environments, as well as adverse weather like extreme heat or cold, rain, snow, and high winds

Be an excellent administrator and have the vision to make long-term plans for a park (superintendents)

Be able to react quickly and effectively in crisis situations (rangers)

Have extensive knowledge about park resources and be able to teach others (interpreters)

A Proud Heritage

In 1872, Congress enacted an historical measure, establishing Yellowstone National Park as "a public park or pleasuring ground for the benefit and enjoyment of people." This landmark act marked the first time our country—or any other country in the world—officially recognized the importance of preserving our most awe-inspiring natural resources in their natural state.

On August 25, 1916, President Woodrow Wilson signed an act creating the National Park Service, a new federal bureau in the Department of the Interior. The mandate of this fledgling bureau was to protect the forty national parks and monuments then in existence and any that would subsequently be created. According to this act, the purpose of these parks was "to conserve the scenery and the natural and historic objects and the wildlife therein and to provide for the enjoyment of the same in such manner and by such means as will leave them unimpaired for the enjoyment of future generations."

An Executive Order in 1933 transferred an additional sixty-three monuments and military sites from the authority of the Forest Service and the War Department to that of the National Park Service. This order laid the cornerstone for the National Park Service we know today, which includes historical sites as well as scenic areas.

Throughout the years, the National Park Service has remained committed to the ideal of conserving our country's natural and cultural resources and has developed a mission statement that reflects this ideal. The mission statement of the National Park Service is as follows: The National Park Service is dedicated to conserving unimpaired the natural and cultural resources and values of the National Park System for the enjoyment, education and inspiration of this and future generations. The Service is also responsible for managing a great variety of national and international programs designed to help extend the benefits of natural and cultural resources conservation and outdoor recreation throughout this country and the world.

Today, the National Park Service includes 374 distinct areas in 49 states, the District of Columbia, American Samoa, Guam, Puerto Rico, Saipan, and the Virgin Islands. Wilson's "Organic Act" of 1916 has since inspired more than one hundred other nations to create similar parks and preserves.

How Do I Become a NPS Employee?

Almost no one enters the National Park Service in the position they would ultimately like to hold. Students who hope to one day serve as a ranger or an interpreter must begin by getting a foot in the door. Most people begin as seasonal employees, working for three to four months a year in parks that receive more visitors during either the summer or winter seasons. This seasonal experience enables people to gain an understanding of the National Park Service mission and determine whether they would enjoy a career within the park system.

Those who choose to continue usually try to get experience in a variety of entry-level positions or in several different parks. This process helps individuals become familiar with the complex park system. It also allows park managers to gauge their strengths and abilities. When a person has gained experience through seasonal positions, he or she may be considered for a permanent

HOW DO I BECOME A . . . ?, CONTINUED

position when one becomes available. Once an individual has gained permanent employment within the park system, he or she will receive extensive on-the-job training. Rangers also undergo fire, search and rescue, and law enforcement training.

Until that first opportunity becomes available, however, there are many ways for individuals to prepare themselves for a career in the National Park Service.

EDUCATION

High School

Students who hope to join the National Park Service should study science and history during high school. They should also stress communication skills. Because interaction with the public is such a significant part of park careers, students also may want to take psychology, education, and sociology courses. Those who plan to become rangers also should concentrate on gym courses; physical fitness is a definite asset for people who must hike miles of backcountry trails, fight fires, and climb rocks to perform rescues.

Hands-on experience can be a distinct advantage for a person who is trying to enter a competitive field. Students who are interested in working for the National Park Service should seek this experience by volunteering for a national park, through the Volunteers in Parks (VIP) program. Park volunteers can help park employees in any number of ways, including answering phone calls, welcoming visitors, maintaining trails, building fences, painting buildings, or picking up litter.

Students who do not live near a national park should contact the Student Conservation Association (SCA), which provides volunteers to assist federal and state natural resource management agencies. The SCA

More Lingo to Learn

National Historic Park: *A national historic park is an area that preserves the location of an event or activity that is important to our country's heritage.*

National Historic Site: *A national historic site is similar to a national historic park, but is usually smaller.*

National Memorial: *These areas commemorate events or individuals of national significance.*

National Monument: *National monuments cover smaller areas than the national parks and do not have as great a diversity of attractions.*

National Park: *National parks cover large areas and contain a variety of resources. Most are chosen for the natural scenic and scientific values.*

National Parkway: *A national parkway is a scenic roadway designed for leisurely driving.*

National Preserve: *A national preserve is an area set aside for the protection of specific natural resources.*

National Recreation Area: *An area or facility that has been set aside for recreational use.*

brings together students from throughout the United States to serve as crew members within the national parks. These students live and work within the parks for four to five weeks at a time.

Both the VIP and SCA experiences can help a student prepare for a career in the National Park Service and determine whether he or she would enjoy such a career.

Postsecondary Training

Though not currently required, prospective park employees would be well advised to obtain a bachelor's degree. Most rangers currently in the park system are college graduates and many believe that this will one day become a requirement. Any individual who hopes to serve as a scientist, archeologist, or historian within the parks must have a college degree, with a major in the relevant discipline. Those who plan to be interpretive rangers should place particular emphasis on science.

Though there is no specific curriculum for people hoping to enter the National Park Service, students should continue to study science, with an emphasis on environmental science. History, public speaking, and business administration courses all would be useful for anyone entering this field.

Because there is so much competition for National Park Service jobs, particularly ranger jobs, many people put themselves through additional training programs to distinguish themselves from other candidates. Some undergo medical technician training programs or police academies. Others attend independent ranger academies to learn the fundamentals of law enforcement, emergency procedures, and fire fighting. These training programs can offer an excellent foundation for a prospective ranger.

CERTIFICATION OR LICENSING

While some park employees, such as architects or attorneys, may need to be certified or licensed in their fields, there is no general certification requirement for park employees. Individuals who become rangers may be given emergency medical training. Those who work in parks with underwater resources, such as the Channel Islands, may become certified divers.

WHO WILL HIRE ME?

Although the National Park Service is the only employer for people who would like to pursue this career, there are many, radically different National Parks.

WHO WILL HIRE ME?, CONTINUED

People who pursue this career may work in mountainous parks like the Grand Tetons, desert parks like Saguaro National Monument, forested parks like Yellowstone, or even marine parks, such as the Channel Islands.

The skills necessary to many positions within the National Park Service also are highly transferable. Interpretive rangers, for instance, may pursue careers as botanists, educators, or naturalists. Law enforcement rangers may consider careers as police officers, fire fighters, or emergency medical personnel. The scientists who study our parks' resources may move into private research, or, like the historians and archaeologists, they may consider becoming educators.

WHERE CAN I GO FROM HERE?

As is true of most professions, advancement within the National Park Service usually means assuming managerial and administrative responsibilities. Rangers, for instance, may become *subdistrict rangers, district rangers,* and then *chief rangers.* Chief rangers may one day become park superintendents. Superintendents, in turn, may assume regional responsibilities. While this is the traditional path to advancement, it is not one that anyone treads very quickly. The opportunities for upward mobility within the National Park Service are limited because the turnover rates at upper levels tend to be quite low. While this may hinder an ambitious employee's advancement, it is indicative of a high level of job satisfaction.

"The opportunities for upward mobility within the National Park Service are limited because the turnover rates at upper levels tend to be quite low."

WHAT ARE SOME RELATED JOBS?

Because there are so many different positions within the National Parks, it is difficult to generalize about the related jobs. Law enforcement rangers may find similar work as police officers, security personnel, fire fighters, or emergency medical professionals. Interpretive work is related to education, botany, and zoology. Only in the National Parks, however, do people have an opportunity to exercise their skills while surrounded by such breathtaking natural beauty.

Related Jobs

Botanists
Emergency medical technicians
Firefighters
Police officers
Security personnel
Teachers
Zoologists

WHAT ARE THE SALARY RANGES?

"My father, who knew how much I loved my job, used to say that I got paid in psychic dollars," says Carol Spears. "He knew I'd never become a millionaire in this job, but he saw that I loved what I was doing."

The salaries for National Park employees are based on individuals' level of responsibility and experience. Employees are assigned salary grade levels. As they gain more experience, they are promoted to higher grade levels, or to higher salary steps within their grade levels.

The National Park Service uses two categories of levels. The first, called the General Schedule (GS), applies to professional, administrative, clerical, and technical employees and is fairly standard throughout the country. Fire fighters and law enforcement are included in the General Schedule. The other, called the Wage Grade (WG), applies to employees who perform trades, crafts, or manual labor and is based on local pay scales.

Most rangers begin at or below the GS-5 level, which, in 1996, translated to earning between $22,233 and $26,303. The average ranger in 1996 was on the second step of the GS-7 level, which translates to a salary of $25,897. The most experienced rangers can earn $32,582, which is the highest salary step in the GS-7 level. To move beyond this level, most rangers must become supervisors, subdistrict rangers, district rangers, or division chiefs, like Carol Spears and Jim Webster. At these higher levels, people can learn up to $70,000 per year. These positions are difficult to obtain, however, because the turnover rate for positions above the GS-7 level is exceptionally low.

WHAT IS THE JOB OUTLOOK?

Although it covers a lot of ground, the National Park Service is really a very small government agency. "There are army bases that employ more people than the whole National Park Service," says Jim Webster.

Because the agency is small, job opportunities are limited and, although they are not highly lucrative, they are considered very desirable among individuals who love outdoor work and nature. Consequently, compe-

WHAT IS THE JOB OUTLOOK?, CONTINUED

tition for National Park Service jobs is very intense. This is not a situation that is likely to improve, since turnover rates are low and new parks are seldom added.

Students who are interested in working for the National Parks should not be discouraged, however. "A determined, hard-working, smart person will always be able to get into the National Park Service if they really want to," says Jim. "They just have to be willing to start at the bottom and work their way up."

Oceanographer

Summary

Definition
Oceanographers *apply scientific principles and procedures to the study of the ocean. They strive to understand the chemical, physical, and geological composition of the ocean, the patterns of life within the ocean, and the relationship between the ocean and the atmosphere.*

Alternative Job Titles
Marine scientist
Ocean scientist

Salary Range
$18,000 to $40,000 to $130,000

Educational Requirements
Bachelor's degree
Master's degree or doctorate strongly recommended

Certification or Licensing
None

Employment Outlook
About as fast as the average

High School Subjects
Biology
Chemistry
Earth science
English (writing/literature)
Foreign language

Personal Interests
Boating
The Environment
Science
Wildlife

Summer has come to the Arabian Ocean, bringing with it intense heat and blazing sunlight. Dr. Ken Brink and a team of five other oceanographers gather at the back of the ship to confer. Though their voices are low, their faces betray the concern and frustration they are feeling.

Ken and his colleagues are gathering data about the Arabian Ocean from aboard the T. G. Thompson, a research ship run by the University of Washington. A temperamental piece of equipment is jeopardizing their project. The instrument, called the SeaSoar, is being towed behind the ship. Loaded with sensors, the SeaSoar enables scientists to measure the depth, temperature, salinity, ambient light, chlorophyll fluorescence and zooplankton abundance at various depths.

Members of Ken's team quickly conclude that they must haul the SeaSoar out of the water and attempt to repair it. As they begin the arduous task of pulling hundreds of meters of cable out of ocean, the team encounters another difficulty. The cable has been damaged. They now must try to repair the cable as they pull it in. An hour's job has just become a day-long project.

Sweating profusely, they continue the painstaking effort. Word of their difficulty has begun to spread to the other science teams aboard the Thompson. Members of these teams gradually move to the back of the ship to assist. Before long, every available scientist and many crew members have joined the effort.

Late that evening, after nearly eight grueling hours, the SeaSoar is pulled onto the deck. Tomorrow, the team will be able to repair it and resume collecting data.

"In some ways this was a really frustrating experience," says Ken, "but in others it demonstrates some of the greatest things about oceanography. Professionals in this field tend to help one another out. It's a very cooperative field."

What Does an Oceanographer Do?

Oceanographers investigate the ocean through scientific study. They explore the physical, chemical, and biological makeup of the seas, the geological structure of the seabed, and the relationship between the oceans and the atmosphere. Because the subject is so broad, however, oceanographers usually concentrate on one particular area of study. There are six primary subspecialties within oceanography: biological oceanography, physical oceanography, chemical oceanography, geological oceanography, ocean engineering, and marine policy.

Biological oceanographers study the many forms of life in the sea. Unlike marine biologists, who study the physiology and habits of individual organisms, biological oceanographers strive to understand the relationship between living organisms and their environment. They study patterns in population density, life cycles, and the cycling of nutrients through the marine food chain. They also examine the distribution of plants and animals through the ocean, the interrelationships between different organisms, and the impact of human behavior on ocean life.

Physical oceanographers examine physical forces and features within the ocean. They observe and record the currents, temperatures, density, salinity, and acoustical characteristics of the ocean. Physical oceanographers are also concerned with the interaction between the ocean and the atmosphere. A

Lingo to Learn

Abyssal plain: *The deep ocean floor, an expanse of low relief at depths of four thousand to six thousand feet.*

Coriolis effect: *The deflection of air or water bodies, relative to the solid earth beneath, as a result of the earth's eastward rotation.*

Ekman circulation: *Movement of surface water at an angle from the wind as a result of the Coriolis effect.*

Epibenthic: *Living on the surface of the bottom of the ocean.*

physical oceanographer might, for instance, try to determine how the ocean influences the climate or weather.

Chemical oceanographers are interested in the chemical characteristics of the ocean and the chemical interactions that occur between the ocean, the atmosphere, and the sea floor. Chemical oceanographers try to identify and understand the forces that determine the ocean's chemical composition and they study the impact this composition has on living organisms and manmade materials. They also investigate ocean resources that may be useful for fuel, food, or medicine and they study the effects of pollution on the ocean.

Geological oceanographers explore the shape and material of the seafloor in order to draw conclusions about the origins of ocean sediment and about the patterns of the ocean's geological features. They map underwater mountains, ridges, and valleys and they study sediment samples from the ocean floor to learn about the history of oceanic circulation and climates. Geological oceanographers attempt to understand the origin of volcanoes and the gradual movement of the earth's surface. They also try to identify potential sources of oil, gas, and minerals.

Oceanography requires sophisticated equipment and instrumentation that has been adapted to the unique underwater environment. The individuals who specialize in designing and building equipment for ocean research are called *oceanographic engineers.* To design oceanography instruments, these engineers must understand research methods and the way materials react to conditions beneath the ocean's surface.

Marine policy experts are oceanographers who use their secondary knowledge of law, business, or the social sciences to help develop responsible policies for use of ocean and coastal resources.

Although individual oceanographers concentrate on specific characteristics of the ocean, the living organisms, chemical composition, physical characteristics, and geological features within the ocean are all interrelated. Consequently, oceanographers from the various subspecialties must work closely together. Many oceanographic research projects are interdisciplinary and involve several ocean scientists with different areas of expertise.

Oceanographers may conduct research from a ship, in a laboratory, or at a desk. While oceanographers spend the majority of their time on land analyzing data, many also go to sea at least once or twice each year. While at sea, oceanographers use weighted hollow tubes, called *corers,* to collect sediment samples. They use cameras equipped with underwater lights to view the depths and echo sounders to measure the distance to the ocean floor. Sonar devices

How Does an Oceanographer Do?, continued

enable the researchers to map the shape and features of the ocean floor. Other devices enable them to measure temperature, density, salinity, magnetic variations, and gravitational pull. One two-week research cruise can provide an oceanographer with enough data to study for an entire year.

Some oceanographers also gather data by observing the ocean's surface from airplanes. Those who specialize in estuaries and coastal waters may be able to dive from small boats to collect samples and make observations. Still other oceanographers are able to collect data without leaving their offices. These scientists obtain data from remote-sensing satellites or by analyzing mathematical models.

What Is It Like to Be an Oceanographer?

Ken Brink is quick to explain that he does not spend the majority of his time at sea or hauling cables out of the ocean. Most of the year, Ken can be found behind his desk at the Woods Hole Oceanographic Institute.

As a senior scientist at Woods Hole, Ken teaches one graduate course every two years. He is also responsible for advising graduate students and overseeing scientists who are working on postdoctoral projects.

For the past year, Ken has served as the chairman of the Ocean Studies Board, an organization affiliated with the National Academy of Science. The Ocean Studies Board is responsible for appointing groups of independent scientists to study matters of environmental policy. The board then evaluates the panels' findings and presents recommendations to the appropriate individuals within government, such as congressional staffers, agency administrators, and members of the National Oceanic and Atmospheric Administration (NOAA). The work is interesting, but arduous, requiring nearly half of Ken's time. "Nearly every day there is something that must be handled immediately," says Ken.

At present, Ken is not able to spend much time on research. When he is not doing work for the Ocean Research Board, working with students, or grading papers, he is usually busy reviewing research proposals and evaluating prospective articles for scientific journals. When he finds time to conduct his own research, he considers it "a rare treat."

"My stint as the chair of the Ocean Research Board will be over soon, however," he says with a smile. "Then I'll have more time for research."

Despite the rigors of his academic and bureaucratic responsibilities, Ken manages to spend a few weeks each year at sea, gathering data. He looks forward to these trips for both the research and the camaraderie. "At sea, you might work like a dog for eighteen straight hours, but when you're done for the day you have a great sense of accomplishment. Oceanographers also have a reputation for working hard and then having a great time."

Like Ken, Dr. Aubrey Anderson is both a teacher and a research scientist. A professor of oceanography at Texas A&M University, Aubrey specializes in using acoustical physics to study the ocean. He estimates that he spends approximately a third of his time teaching graduate courses, a third conducting research, and a third performing administrative tasks.

"I enjoy the teaching and I look forward to being in class," says Aubrey. "I enjoy the interaction with students. It's very gratifying to see them grow as individuals. In some ways, it's like having a lot of children."

Aubrey also enjoys conducting research and looks forward to his seafaring excursions. "Being at sea can either be very boring or it can be like a fire drill. More often, though, it's like a fire drill. It can cost anywhere from $8,000 to $50,000 a day to take a research vessel out, so you have to make the most of your time. Once the ship reaches the first data collection station, you go to work—no matter what the weather's like and no matter what time of day it is—and you work until you finish at that station. You don't just work on your project, either. There are usually several scientific teams on a research ship and everyone helps everyone else."

Aubrey readily confesses that he does not enjoy the administrative responsibilities of his position. He explains, "Administrative problems are the same no matter where you go. There isn't enough money, there isn't enough space, and there aren't enough resources to go around. Human beings are the same everywhere; they tend to argue about petty things."

Both Ken Brink and Aubrey Anderson feel that oceanography is an exciting, deeply rewarding profession. "You look at some problem that you don't understand and then one day you can say 'Aha! This is why that happens,'" says Ken. "When that happens, it's wonderful."

Have I Got What It Takes to Be an Oceanographer?

To succeed as an oceanographer, a person must be intelligent, willing to work hard and at irregular hours, and able to work closely with other people. Oceanographers must have superior computer and math skills. They must also be able to tolerate sea travel; people who suffer from seasickness may want to think twice about choosing a career in seagoing oceanography!

HAVE I GOT WHAT IT TAKES . . . ?, CONTINUED

Cooperation is particularly important in this field, since oceanographers must work closely together on research projects. Ken explains, "Oceanographers regularly spend weeks or months together on a boat. Personalities with a lot of rough edges just aren't going to make it. I think seagoing oceanography just naturally tends to select for calmer personalities."

Because oceanographers must publish the results of their research, excellent writing skills are also essential. "I originally got into science because I hated writing," Ken recalls ruefully. "When I got here I found out that writing is a big part of the job."

The most important characteristic for an oceanographer, however, may be intellectual curiosity. Oceanographers must yearn to solve nature's mysteries. Aubrey Anderson, for instance, vividly recalls the day he and a research team pulled a core sample out of the Gulf of Mexico only to have the sediment sample shoot out of the corer like a cannon ball. "It was obviously pushed out by pressurized gas," he said, "but we all knew it was impossible for the gas bubbles at the bottom of the ocean to be big enough to shoot a core sample across the deck. When it happened a few more times, we figured out that the gas existed in a solid form, like ice, at the bottom of the ocean. We were very excited to discover gas hydrate ice could exist in the Gulf of Mexico.

"Of course, we also learned pretty quickly not to stand directly in front of the corer when it was pulled out of the water!"

HOW DO I BECOME AN OCEANOGRAPHER?

EDUCATION

High School

Students who are interested in a career in oceanography should take as many math, science, and computer courses as possible. English and communications courses are also helpful. Because oceanographers from all over the world work together and share information, language classes also are recommended.

Ken Brink advises high school students to learn Spanish, Russian, Chinese, or Japanese. "Most of the oceanographers in Western Europe speak English, whereas relatively few people in Latin America, Asia, or Eastern Europe speak English. I was at a conference in Ukraine last year where only about 20 percent of the scientists in attendance spoke English. It would have been very helpful to be able to speak Russian."

Students who are interested in this career should seek opportunities to gain firsthand oceanographic experience. Many coastal universities offer summer camp programs that enable young people to collect and analyze ocean data. Sea Grant, a federally funded program available in most states, also offers students oceanographic experiences.

Postsecondary Training

Once in college, prospective oceanographers should continue to take science courses, including biology, chemistry, physics, and geology. While some universities do offer undergraduate oceanography programs, students who plan to go on to graduate school should not necessarily major in marine science or oceanography. In fact, most oceanographers concentrate on a related area of science, such as chemistry, physics, geology, or biology, before studying oceanography in graduate school. A well-rounded background in science is essential to a career as an oceanographer.

"A solid background in the sciences can be applied in a lot of different ways. You don't have to make a final decision right away," says Ken Brink, who majored in applied physics as an undergraduate student. "I studied oceanography in graduate school because it offered me an opportunity to study the ocean and to do some serious physics at the same time."

Students who plan to pursue an advanced degree in oceanography should look for institutions that offer significant hands-on research experience. Approximately fifty universities offer graduate programs in oceanography. Ten of the largest U.S. oceanographic institutions form the Joint Oceanographic Institutions, Inc. These institutions are: University of Hawaii, Lamont-Doherty Geological Observatory of Columbia University, University of Miami, Oregon State University, University of Rhode Island, Texas A&M University, The University of Texas Institute for Geophysics, University of Washington, Scripps Institution of Oceanography of the University of California, and Woods Hole Oceanographic Institution. Most of these institutions offer programs in each of the subfields of oceanography, including ocean engineering.

WHO WILL HIRE ME?

At present, the government employs approximately 50 percent of practicing oceanographers. Another 40 percent hold academic positions. The remainder work for private industry and not-for-profit environmental organizations.

WHO WILL HIRE ME?, CONTINUED

Oceanographers who work for the government may be responsible for resource management, research and development, environmental monitoring and protection, or regulation enforcement. Within the federal government, oceanographers are employed by the Department of Energy, Minerals Management Service, U.S. Geological Survey, National Oceanic and Atmospheric Administration, Naval Oceanographic Office, Naval Research Laboratory, Office of Naval Research, and National Science Foundation.

Academic positions usually entail a combination of teaching and research. Experienced oceanographers who work for large universities may devote the lion's share of their time to research, teaching only one or two classes each year. Oceanographers who teach at smaller institutions or at the undergraduate level, on the other hand, may be entirely occupied with teaching. Within academia, the two largest employers of oceanographers are Woods Hole Oceanographic Institution and the Scripps Institution of Oceanography, though many coastal universities also maintain excellent oceanographic programs.

The Challenger Expedition 1872-1876

Oceanography, the systematic exploration and study of the world's oceans, did not begin until December 7, 1872, when the HMS Challenger set sail from Sheernes, England. Throughout the ensuing expedition, which was to last more than 4 years and cover 68,890 nautical miles, Wyville Thomson and six other scientists carefully measured and recorded information about depth, temperature, currents, and contours of the great ocean basins.

Thomson, a professor of natural history at Edinburgh University, had begun planning the unprecedented adventure in 1870. With help from the Royal Society of London, he was able to persuade the British government to provide one of Her Majesty's ships for the voyage. The ship, a wooden corvette that weighed 2,306 tons, was essentially a sailing vessel, although it did possess an engine for maneuvering.

Between December 1872 and May 1876, the Challenger traversed the Atlantic Ocean several times, sailed through the Pacific Ocean, visited the polar seas, and crossed the Antarctic circle. Thomson and his colleagues recorded measurements and gathered samples at 362 stations. At each station, the scientists measured the depth of the ocean and its bottom temperature, collected samples of the bottom sediment and water, collected samples of fauna from various depths, recorded the direction and rate of the surface current, and noted atmospheric conditions. In the process, Challenger scientists discovered 4,717 new species.

The results of the Challenger Expedition were published between 1885 and 1895, in a fifty-volume report that is still used by oceanographers today. In 1895, John Murray, a member of the Challenger Expedition, described this report as "the greatest advance in knowledge of our planet since the celebrated discoveries of the fifteenth and sixteenth centuries."

Private sector oceanographers may be engaged in research and development or resource management. A geological oceanographer, for instance, might help a petroleum company locate new sources of petroleum beneath the ocean floor. A biological oceanographer might work for a pharmaceutical company, trying to identify chemicals that could lead to the development of new medicines. A manufacturer might employ an oceanographer to ensure that its plants are not violating federal or state regulations by polluting coastal waters.

The insurance industry is a recent, but growing source of private sector career opportunities for oceanographers. Judi Rhodes, executive director of the Oceanography Society, notes, "Insurance underwriters are concerned about global warming and what it will mean to human life. A relatively small increase in global temperature could lead to a significant rise in the sea level. The loss liability is enormous. Insurance underwriters' concerns have created a new niche for marine scientists and oceanographers."

Judi also encourages students interested in oceanography to consider combining oceanographic knowledge with expertise in another field, such as communications or graphic arts. "Science writing," she explains, "is sadly lacking. This lack of communication contributes to less-than-adequate funding. Students who are good communicators will benefit the entire field. Those who acquire a second skill set also are able to offer employers an extra dimension and are more marketable."

While positions in private industry tend to offer higher compensation than academic or governmental positions, industry scientists are expected to study topics of concern to their employer. Oceanographers who work for universities usually have more freedom to pursue the questions and ideas that interest them.

Advancement possibilities

Marine educators *teach marine science at aquariums, museums, colleges, and universities.*

Marine and ocean engineers *use their scientific and technical knowledge to design and build instruments and machinery that assist oceanographers in their research.*

Marine policy experts *are oceanographers with extensive training in the social sciences, law, or business. They use this knowledge to help develop guidelines and policies for wise use of ocean resources.*

Where Can I Go from Here?

People who bring only a bachelor's degree or a master's degree to the study of oceans will find themselves competing with oceanography graduate students

WHERE CAN I GO FROM HERE?, CONTINUED

for a limited number of technical assistant positions. The work can be interesting, often requiring sojourns at sea, but the opportunities for advancement are limited. Individuals who are not interested in conducting oceanographic research, however, may build rewarding careers designing or selling oceanographic equipment, advising industry, or developing communications materials for environmental organizations. Such positions may lead to opportunities in middle or upper management.

Most research and teaching positions are reserved for oceanographers with doctoral degrees. Oceanographers who complete a doctorate usually begin their careers in postdoctoral positions, assisting experienced oceanographers with research projects. After gaining two or three years of experience, an individual may become an assistant professor or assistant scientist. As such, the oceanographer may continue to assist more experienced scientists, and also assume teaching responsibilities. After establishing credentials as a knowledgeable, creative scientist, an oceanographer may begin to conduct his or her own research projects.

Experienced oceanographers usually can receive promotions and earn higher salaries by assuming administrative responsibilities. Administrative posts are not for everyone, however. Ken Brink, for instance, says, "I don't want to become a department head or a dean. These options are available, but all I want is to do good research and to help others do good research."

WHAT ARE SOME RELATED JOBS?

The U.S. Department of Labor classifies oceanographers under theoretical research (GOE) and geology occupations (DOT). Also included in these categories are people who conduct research to learn about the Earth, climate, atmosphere, chemical composition of our environment, and about living organisms. Some related jobs include astronomers, meteorologists, chemists, geographers, geodesists, geologists, petrologists, mineralogists, hydrologists, mathematicians, paleontologists, geophysicists, and physicists.

WHAT ARE THE SALARY RANGES?

When asked about salary ranges, Aubrey Anderson laughs and responds, "Well, I don't know any rich oceanographers."

Nonetheless, for individuals who have excellent academic credentials and the motivation to conduct innovative, thorough research, oceanography can be a financially rewarding and intellectually fulfilling career.

Individuals who enter the field of oceanography with a bachelor's or associate's degree are usually employed as technicians or technical assistants and can expect to earn an annual salary of $18,000 to $20,000. Those with master's degrees in oceanography may be able to earn $30,000 at the outset but, again, their opportunities for advancement and earning potential are limited.

Oceanographers who begin their careers with doctoral degrees can expect to earn $35,000 to $40,000 to start. After a few years, oceanographers may be qualified for associate professorship positions, which typically offer $40,000 to $50,000 per year. Oceanographers with considerable experience may earn anywhere from $70,000 to $130,000, depending on their place of employment and the nature of their work.

Established oceanographers can augment their income by accepting speaking engagements and providing consulting services to not-for-profit and private sector organizations.

WHAT IS THE JOB OUTLOOK?

A variety of forces are at work in shaping the outlook for oceanography. In recent years, the federal government has reduced funding for some areas of scientific research. Because so many universities rely on public funding, the reduction in public funds has curtailed academic research, reduced the number of available student grants, and eliminated some research and teaching positions. This is particularly true of oceanography, which, because it requires sea excursions, is a particularly expensive science to practice.

At the same time, students continue to enter oceanography graduate programs. A renewed emphasis on environmental concerns, the popularity of Jacques Cousteau, and the increased awareness of sea mammals have all contributed to the public's interest in oceanography. Coupled with the reduction of public funds, this interest has made oceanography a highly competitive field.

On the other hand, heightened public awareness of environmental issues has led to the creation of new private sector positions for earth scientists. In response to pressure from consumers, private industry has begun employing scientists, including oceanographers, to oversee their use of environmental resources and their waste disposal procedures.

WHAT IS THE JOB OUTLOOK?, CONTINUED

While the increase in private sector jobs is not enough to counteract the dearth of academic positions caused by the reduction in public funds, the future may hold more promise. "Public awareness of environmental concerns is on the upswing," says Judi Rhodes of the Oceanography Society. "Since constituency awareness and concern are supposed to drive federal spending, it is possible that we will see increased resources in the future."

"I would still encourage a bright student to pursue an oceanographic career," concludes Ken Brink, "because it's an exciting field with a lot of good problems still to be solved. And no matter how tough the field becomes, there is always room for extremely talented people."

Jacques-Yves Cousteau: Oceanographer and Environmentalist

Jacques Cousteau, the famed oceanographer and champion of the environment, may have made his greatest contributions in his capacity as a teacher. Not content to share his discoveries with fellow oceanographers and the scientific elite, Cousteau made it his life's mission to share the wonders of the depths with people from all walks of life. In the process, he may have done more than any other individual to alert the world to the importance and necessity of preserving our oceans' resources.

Jacques-Yves Cousteau was born on June 11, 1910, in St. André de Cubzac, a small town near Bordeaux, France. At 20, he entered the French Naval Academy and began training to become a naval pilot. When a near-fatal automobile accident put an end to his aviation career in 1936, Cousteau transferred to sea duty and discovered a love for the ocean that was to last his lifetime.

For the next fifty years, Cousteau shared his many underwater adventures and discoveries with the world through more than seventy films, more than fifty books, and the immensely popular television series, "The Undersea World of Jacques Cousteau."

A self-described oceanographic technician, Cousteau also developed numerous tools for exploring the ocean depths. In 1943, Cousteau and Emile Gagnan developed the first regulated compressed-air breathing device for sustained diving. In 1959, he and Jean Mollard designed a maneuverable diving saucer capable of carrying two people to a depth of 350 meters. In 1982, he and two colleagues developed the Turbosail wind-propulsion system.

Throughout his career, Cousteau was a passionate environmental advocate drawing attention to the dangers of pollution, overpopulation, and the mindless squandering of the world's natural resources. "Future generations," he said, "would not forgive us for having deliberately spoiled their last opportunity and the last opportunity is today."

For his efforts, Cousteau received the United Nation's International Environmental Prize and the U.S. Presidential Medal of Freedom. In 1974, he founded The Cousteau Society, which is dedicated to protecting and improving the quality of life for present and future generations.

When Jacques Cousteau died in 1997, at the age of eighty-seven, he was mourned by people around the world. Said French President Jacques Chirac, "[Cousteau] represented the defense of nature, modern adventure, and the invention of the impossible."

Section 3
What Can I Do Right Now?
environment

Get Involved: A Directory of Camps, Programs, Competitions, Etc.

First

Now that you've read about some of the different careers available in the environmental field, you may be anxious to experience this line of work for yourself, to find out what it's *really* like. Or perhaps you already feel certain that this is the career path for you and want to get started on it right away. Whichever is the case, this section is for you! There are plenty of things you can do right now to learn about environmental careers while gaining valuable experience. Just as important, you'll get to meet new friends and see new places, too.

In the following pages you will find more than fifty programs run by organizations that share your commitment to the environment, and all of them can help you turn your interest into a career. Some organizations offer just one kind of program: colleges, quite naturally, will probably offer only academic courses of study. Other organizations, such as wildlife sanctuaries, may offer internships and job opportunities in addition to summer camps and courses. It's up to you to decide whether you're interested in one particular type of program or are open to a number of possibilities. The types of programs available are listed right after the name of the program or organization, so you can skim through to find the listings that interest you most.

THE CATEGORIES

Camps

When you see an activity that is classified as a camp, don't automatically start packing your tent and mosquito repellent. Particularly where academic study is involved, the term "camp" sometimes simply means a residential program including both educational and recreational activities. However, when studying the environment, even some academic work must take place outdoors. That means that many of the camps in this volume do include actual camping, so you may need that tent after all. Since there are quite a few camps of both kinds listed here, be sure to read the descriptions thoroughly so you can tell the difference.

College Courses/Summer Study

These terms are linked because most college courses offered to students your age must take place in the summer, when you are out of school. Similarly, many summer study programs are sponsored by colleges and universities that want to attract future students and give them a head start in higher education. Summer study of almost any type is a good idea because it keeps your mind and your study skills sharp over the long vacation. Summer study at a college offers any number of additional benefits, including giving you the tools to make a well-informed decision about your future academic career. We have included many study options in these listings, including some outstanding college and university programs.

Competitions

Competitions are fairly self-explanatory, but you should know that there are only a few in this book because environmental competitions on a regional or national level are relatively rare. In your own city or county, however, you should be able to find some civic groups or government agencies that sponsor essay contests, recycling drives, and other competitions related to your interest in the environment.

Conferences

Conferences for high school students are usually difficult to track down, since most are for professionals in the field who gather to share new information and ideas with each other. Don't be discouraged, though. A number of professional organizations with student branches invite those student members to their conferences and plan special events for them. Some student branches even run their own conferences; check the directory of student organizations at the end of this book for possible leads. This is an option worth pursuing because conferences focus on some of the most current information available and also give you the chance to meet professionals who can answer your questions and even offer advice.

Employment Opportunities

As you may already know from experience, employment opportunities for teenagers can be very limited. This is particularly true in environmental professions, which require workers with bachelor's and even graduate degrees in addition to research experience. There *are* a few jobs in the field for high school students, but you may just have to earn your money by working at a mall or restaurant and get your environmental experience in an unpaid position elsewhere. Bear in mind that, if you do a good enough job and the group you work for has the funding, this summer's volunteer position could be next summer's job.

Field Experience

This is something of a catchall category for activities that don't exactly fit the other descriptions. But anything called a field experience in this book is always a good opportunity to get out into the natural world and explore the work of environmental professionals. A number of the field experiences listed here are perfect for students who want to explore environmental studies a bit more before making any kind of career decision.

Internships

Basically, an internship combines the responsibilities of a job (strict schedules, pressing duties, and usually written evaluations by your supervisor) with the uncertainties of a volunteer position (no wages or fringe benefits, no guarantee of future employment). That may not sound very enticing, but completing an internship is a great way to prove your maturity, your commitment to the environment, and your knowledge and skills to colleges, potential employers, and yourself. Some internships here are just formalized volunteer positions; others offer unique responsibilities and opportunities. Choose the kind that works best for you!

Memberships

When an organization is in this category, it simply means that you are welcome to pay your dues and become a card-carrying member. Formally joining any organization has the benefits of meeting others who share your interests and concerns, finding opportunities to take action, and keeping up with current events in the field. Depending on how active you are, the contacts you make and experiences you gain may help when the time comes to apply to colleges or look for a job.

In some organizations, you may pay a special student rate but receive virtually the same benefits as a regular adult member. Other groups have student branches with special activities and publications. Still other membership organizations are keen to help you start your own local branch. If you and your friends are at all interested in starting an environmental club, it's a fine idea to associate yourselves with a group that's already established.

Finally, don't let membership dues discourage you from contacting any of these organizations. Most charge no more than about $25 because they know that students are perpetually short of funds. If even that is too much for you, contact the group that interests you anyway—they are likely to at least send you some information and place you on their mailing list.

Seminars

Like conferences, seminars are often classes or informative gatherings for those already working in the field, and are generally sponsored by professional organizations. This means that there aren't all that many seminars for young people. But also like conferences, they are often open to affiliated student members. Check with various organizations to see what kind of seminars they offer and if there is some way you can attend.

Volunteer Programs

Volunteerism is now enjoying great popularity, particularly among young people. Whether you're volunteering to meet your school's community service requirements or simply to help others and support a worthy cause, you can use the experience to explore environmental careers. Caring for wildlife, campaigning for pollution control, and preserving the wilderness are just a few common volunteer activities—the listings in this book and your own ingenuity can lead to many more. Depending on your needs and interests, volunteering can be a long- or short-term commitment, perhaps part-time during the school year or full-time during the summer. This is an option that is flexible and broad enough for almost everyone.

PROGRAM DESCRIPTIONS

Once you've started to look at the individual listings themselves, you'll find that they contain a lot of information. Naturally, there is a general program overview, but wherever possible we also have included the following details.

Application Information
Each listing notes how far in advance you'll need to apply for the program or position, but the simple rule is to apply as far in advance as possible. This ensures that you won't miss out on a great opportunity simply because other people got there ahead of you. It also means that you will get a timely decision on your application, so if you are not accepted, you'll still have some time to apply elsewhere. As for the things that make up your application—essays, recommendations, etc.—we've tried to tell you what's involved, but be sure to contact the program about specific requirements before you submit anything.

Background Information
This includes such information as the date the program was established, the name of the organization that is sponsoring it financially, and the faculty and staff who will be there for you. This can help you—and your family—gauge the quality and reliability of the program.

Classes and Activities
Classes and activities change from year to year, but knowing that a precollege program usually offers "Introduction to Biology" or that a camp generally trains its students in leave-no-trace camping can help you decide if it's the right kind of program for you.

Contact Information
Wherever possible, we have given the *title* of the person you should contact instead of the *name* because people change jobs so frequently. If no title is given and you are telephoning an organization, simply tell the person who answers the phone the name of the program or position that interests you and he or she will forward your call. If you are writing, include the line "Attention: Summer Study Program" (or whatever is appropriate after "Attention") somewhere on the envelope. This will help to ensure that your letter goes to the person in charge of that program.

Credit
Where academic programs (and a few internships) are concerned, we sometimes note that high school or college credit is available to those who have completed them. This means that the program can count toward your high school diploma or a future college degree just like a regular course. Obviously,

this can be very useful, but it's important to note that rules about accepting such credit vary from school to school. Before you commit to a program offering high school credit, check with your guidance counselor to see if it is acceptable to your school. As for programs offering college credit, check with your chosen college (if you have one) to see if they will accept it.

Eligibility and Qualifications

The main eligibility requirement to be concerned about is age or grade in school. A term frequently used in relation to grade level is "rising," as in "rising senior": someone who will be a senior when the next school year begins. This is especially important where summer programs are concerned. Some organizations base admissions decisions partly on GPA, class rank, and standardized test scores. This is mentioned in the listings, but you must contact the program for specific numbers. If you are worried that your GPA or your ACT scores, for example, aren't good enough, don't let them stop you from applying to programs that consider such things in the admissions process. Often, a fine essay or even an example of your dedication and eagerness can compensate for statistical weaknesses.

Facilities

Of course you need to know if you'll be roughing it out in the woods or relaxing in a new dormitory on a college campus. But you should also know about the laboratory facilities, the sailing vessels for marine research and oceanography programs, and other kinds of facilities you'll be using. These can be particularly important considerations if modern equipment and cutting-edge technology play a large part in the career that interests you.

Financial Details

You'll definitely want to know if you'll be paying or getting paid, and just how much money is involved! Prices do tend to go up each year, but they should be close to the figures we quote. We always try to note where financial aid is available, but most academic programs in particular will do their best to ensure that a shortage of funds does not stop you from participating.

Residential vs. Commuter Options

Simply put, some programs prefer that attendees live with other participants and staff members, others do not, and still others leave the decision entirely to the students themselves. As a rule, residential programs are suitable for young people who live out of town or even out of state, as well as for local residents. Commuter programs may be viable only if you live near the program site or if you can stay with relatives who do. Bear in mind that for residential programs especially, the travel between your home and the location of the activity is almost always your responsibility and can significantly increase the cost of participation.

FINALLY . . .

Ultimately, there are three important things to bear in mind concerning all of the programs listed in this volume. The first is that things change. Staff members come and go, funding is increased or withdrawn, supply and demand determine which programs continue and which terminate. Dates, times, and costs vary widely because of a number of factors. Because of this, the information we give you, although as current and detailed as possible, is just not enough on which to base your final decision. If you are interested in a program, you simply must write, call, fax, or email the organization in charge to get the latest and most complete information available. This has the added benefit of putting you in touch with someone who can deal with your individual questions and problems.

Another important point is that we do not actually recommend or endorse any of the programs and organizations mentioned in this book. We simply pass on to you the information they gave us. You'll need to do some research and make your own evaluations. After all, only you can decide which opportunity is the right one for you.

The third thing to bear in mind is that the programs listed here are just the tip of the iceberg. No book can possibly cover all of the opportunities that are available to you—partly because they are so numerous and are constantly coming and going, but partly because some are waiting to be discovered. For instance, you may be very interested in a volunteer program at a state park several hundred miles away from you. Instead of complaining about the distance, why not go to the state park nearest you and ask if you could volunteer there? They may already have a program for you to join or they may start one because you were the first person to show an interest. Or perhaps you would like to take a college course but don't see the college that interests you in the listings. Call their Admissions Office! Even if they don't have a special program for high school students, they might be able to make some kind of arrangements for you to visit or sit in on a class. Use the ideas behind these listings and take the initiative to turn them into opportunities!

THE PROGRAMS

ALLEY POND ENVIRONMENTAL CENTER

Internship, Volunteer Program

The Alley Pond Environmental Center (APEC) is a nonprofit corporation with a limited number of internship openings. Students ages fourteen years and older can participate in the center's work in nature and science education and also help maintain the center's facilities and surrounding area. As an intern, you might assist with the upkeep of APEC's mini-zoo, classroom areas, and nature trails.

The Alley Pond Environmental Center also offers in-school programs and presentations for high school students. Contact APEC for more specific details on the obligations and benefits of an internship or for information about a more informal position as a volunteer.

Alley Pond Environmental Center
228-06 Northern Boulevard
Douglaston, NY 11363-1890
Tel: 718-229-4000
Fax: 718-229-0376

APPALACHIAN MOUNTAIN CLUB TRAILS PROGRAM

Field Experience, Volunteer Program

The Appalachian Mountain Club (AMC) is a nonprofit organization that relies on volunteers to help accomplish its mission of building and maintaining hiking trails, skiing trails, and trail shelters at various locations in New Hampshire, Connecticut, and Massachusetts. Each summer, the AMC assembles trail crews of volunteers of similar ages and ability levels. Young people between the ages of sixteen and nineteen can join a Classic Teen Crew, Spike Teen Crew, or Advanced Teen Trail Crew operating from the Camp Dodge Volunteer Center at the White Mountain National Forest in New Hampshire. On the Classic Teen Crew, you spend three days working out of Camp Dodge's comfortable accommodations and two days camping out in the backcountry. Once you have worked on the Classic Teen Crew, you can volunteer for the Spike Teen Crew, which spends the entire week camping and working in a remote location. The Advanced Teen Trail Crew is for very experienced volunteers who want to enjoy a weekend of hiking and rock climbing between two weeks of camping in the backcountry and working on the trails. You can also participate in a Classic Teen Crew based out of the Mount Greylock Ski Club at Goodell Hollow in the Berkshires. All accommodations, tools, training, and meals are included in the participation fees; it costs $75 to join the week-long crew and $275 for the two-week option. There is a slight reduction in fees for AMC members. Participants are responsible for their own transportation to and from the main rendezvous point. There is no formal application deadline, but interested teenagers should apply early in the year because space is limited.

Additionally, the Appalachian Mountain Club offers many volunteer opportunities for interested people ages sixteen and older. All participants are expected to perform demanding physical tasks and so should be in good health. For more information or to apply, contact the Volunteer Coordinator.

Appalachian Mountain Club Trails Program
PO Box 298
Gorham, NH 03581
Tel: 603-466-2721
Fax: 603-466-2822

AQUATIC STUDIES CAMP AT SOUTHWEST TEXAS STATE UNIVERSITY

Camp

The Aquatic Studies Camp is run by the Edwards Aquifer Research and Data Center at Southwest Texas State University. It offers students ages thirteen to fifteen the chance to explore and conduct research on various bodies of water, from artesian wells to ponds to rivers. During the week-long residential camp, your days start early (around 6:45 AM) and include snorkeling and diving lessons, lectures, videos, and evening recreational activities. You go on many field trips, including a rafting adventure and a behind-the-scenes visit to Sea World. There are three regular sessions each summer, and a special session with advanced academic studies for eleven- to fifteen-year-olds. Each session costs about $450, which includes room, board, equipment, and field trips. Twenty-six students are accepted into each session on a first come, first served basis. All participants should be accustomed to moderate physical activity and be comfortable around water, preferably with the ability to swim.

The Aquatic Studies Camp also provides regular sessions for nine- to eleven-year-olds and eleven- to thirteen-year-olds.

Aquatic Studies Camp at Southwest Texas State University
Southwest Texas State University, EARDC
248 Freeman Building
San Marcos, TX 78666-4616
Tel: 512-245-2329

AUSTIN NATURE AND SCIENCE CENTER

Volunteer Program

The Austin Nature and Science Center has opportunities for volunteers ages thirteen and older to greet visitors and help them make the most of the center's facilities. As a volunteer, you rotate between different areas each day. One day you might handle a friendly reptile and introduce it to visitors, explaining its habitat and dietary needs; another day you might lead a tour along the Eco-Detective Trail, helping visitors make discoveries about the ecosystem of the center's pond. Two days of training are provided so that volunteers are knowledgeable and safety-conscious; volunteers are also given additional information about specific animals and plants on the days they work with them. To volunteer at the Austin Nature and Science Center, you should be available on a flexible schedule for at least half of the summer. Most volunteers work either Monday and Wednesday or Tuesday and Thursday for about three hours per day. The Austin Nature and Science Center also offers a Counselor-in-Training (CIT) program for students ages thirteen to sixteen who would like to be assistant counselors at the center's summer camps for younger children. The CIT program includes two days of training and three weeks of probationary volunteer work at the camps. Contact the Public Programs Coordinator for further information.

Austin Nature and Science Center
301 Nature Center Drive
Austin, TX 78746
Tel: 512-327-8181
Fax: 512-327-8745

BENJAMIN E. MAYS SCHOLARS IN BIOLOGY PROGRAM AT BATES COLLEGE

College Course/Summer Study

Bates College offers the Benjamin E. Mays Scholars in Biology Program each July for a select group of rising juniors and seniors. Only eighteen students are selected each year to participate in this twelve-day residential program. The program is usually broken down into five two-day units, with each unit taught by college faculty and focusing on an area of biology important in terms of current techniques and technologies. Units cover biological, chemical, and ecological approaches to a range of topics, all of which you explore through a combination of laboratory and field work. Laboratory work is completed in the modern facilities of the Carnegie Science Hall. The topics of each unit vary from year to year, so be sure to write for current information. Participants live in Bates College's modern dormitories and eat in the Commons; you also have access to the campus' many recreational facilities. All tuition, room and board, and travel costs are paid by Bates College; students must pay only for incidental expenses. Travel must be arranged through Bates College's Campus Travel. Admission is, of course, very competitive and your application form must be accompanied by a letter of recommendation, a high school transcript, and an essay detailing why you are applying. Applications must be received by late April. For further information, contact the Office of Special Projects and Summer Programs at Bates College.

Benjamin E. Mays Scholars in Biology Program at Bates College
Office of Special Projects & Summer Programs
Bates College
163 Wood Street
Lewiston, ME 04240-6016

BUREAU OF LAND MANAGEMENT—MONTANA STATE OFFICE

Employment Opportunity, Internship, Volunteer Program

The Bureau of Land Management (BLM) is the division of the U.S. Department of the Interior responsible for managing the land and resources of over 270 million acres of public-owned property. The BLM for Montana and the Dakotas runs the Student Volunteer Service Program for young people who are attending an accredited high school or trade school at least part-time. As a volunteer in the program, you can take part in research projects, help to monitor wilderness areas, perform office work, or participate in any number of activities designed to conserve and preserve the land's natural and historic resources. Depending on your interests and the BLM's needs, you can work alone or in a group, on a prearranged project or on one of your own devising, for a few hours or every day, just once or on a continuing basis. Before beginning work, you will draw up a formal volunteer agreement and the BLM will decide whether you require any special training (including safety training) from them. Upon completing the terms of your agreement, you receive a formal record of your accomplishments as a volunteer and a certificate of appreciation. It is possible to arrange an internship for academic credit if you work with the BLM and your guidance counselor to set up the terms of the volunteer agreement. The BLM for Montana and the Dakotas also participates in the Department of the Interior's Student Educational Employment Program (SEEP). High school students in SEEP are employed at entry-level positions with the BLM that match their interests and career goals. They must be recommended by their school, be in good academic standing, and willing to complete 640 hours (sixteen full weeks) of work before graduation. Successful completion of SEEP may lead to permanent opportunities in federal service upon completion of other educational requirements (namely, a college degree).

College students at the undergraduate and graduate levels are also eligible to participate in SEEP. The Bureau of Land Management welcomes volunteers of all ages and skill levels. For further information, contact the BLM office nearest you: the Montana and Dakotas branch has four district offices in addition to the state office listed here.

Bureau of Land Management—Montana State Office
222 North 32nd Street
PO Box 36800
Billings, MT 59107-6800
Tel: 406-255-2827

CANYONLANDS FIELD INSTITUTE

Camp, College Course/Summer Study, Field Experience

The Canyonlands Field Institute, a nonprofit organization in its thirteenth year, offers the chance to learn how to enjoy and care for the environment at the Colorado Plateau. Students ages thirteen to eighteen can participate in the Whitewater Academy for Teens, which involves rafting on the Colorado River for beginning boaters. The program emphasizes all the skills necessary for river-running, such as navigation, maneuvering techniques, and handling equipment. You also learn minimum-impact camping techniques. Paddling the Goosenecks is a program for fourteen- to eighteen-year-olds keen to take an inflatable kayak trip on the San Juan River. This program offers further instruction in boating and outdoor skills as well as the investigation of wildlife unique to the Colorado Plateau. There is also an expedition to Arches National Park. Both programs last one week each and are offered only once per summer. Only thirteen to twenty people are accepted into each session; early registration is advised, but there is no formal deadline. Most gear and meals are provided in the tuition fees that range from $500 to $700; some financial assistance is available, and members of the Canyonlands Field Institute receive a discount. Staff are licensed whitewater guides with college degrees who are certified in providing emergency medical care.

The Canyonlands Field Institute also offers a number of study and adventure opportunities for school groups, college students (with credit possible), and adults. Office staff will handle all questions and applications.

Canyonlands Field Institute
PO Box 68
Moab, UT 84532
Tel: 800-860-5262
Fax: 801-259-2335
Web: http://www.canyonlandsfieldinst.org

CAROLINA RAPTOR CENTER

Internship

The Carolina Raptor Center (CRC) has been working since 1981 to care for injured and orphaned birds of prey (raptors) while running an environmental education program for visitors to the center and for local schoolchildren. Volunteers are largely responsible for the continuing success and growth enjoyed by the CRC, and high school students are encouraged to donate their time. Opportunities are available in the educational, animal care, administrative, and fund-raising areas of operation. Work may include training birds, preparing exhibits at the center, maintaining the nature trail and its aviaries, or working in the office or gift shop. All volunteers, but especially those working directly with the raptors, are given comprehensive training by experienced staff members. Students who wish to make a long-term commitment to the CRC can apply for an internship. Upon completion of a predetermined number of hours and with a satisfactory evaluation by the staff, interns can receive high school or college credit if their school has given its approval. Because of the amount of training and supervision involved, the CRC can accept only a limited number of interns per season, so students are invited to apply well in advance of their anticipated start date. Call the Carolina Raptor Center for more information about volunteering or interning, or take advantage of its year-round opening and visit the CRC to find out more about its work.

Carolina Raptor Center
PO Box 16443
Charlotte, NC 28297
Tel: 704-875-6521

CATALINA ISLAND SEA CAMP

Camp, Field Experience

Catalina Island Sea Camp features a scuba diving camp on Santa Catalina, an island twenty-six miles long off the coast of southern California. You must be thirteen or older to attend. Campers become certified scuba divers during the three-week course. In addition to scuba certification, you also learn the basics of marine biology, oceanography, and island ecology. You explore tide pools and marine life of the Catalina Island system—sharks, skaters, rays, and marine mammals. A typical day consists of scuba diving in the morning and two-and-a-half hours of labs and instruction in the afternoon. Catalina Island campers partake of island life in their spare time. You can learn to sail, sea kayak, snorkel, rock climb, and shoot underwater photography. During the camp session, everyone takes a day hike to the beach located on the other side of the island. All instructors at the Catalina Island Sea Camp have degrees in marine science, and scuba diving instructors are fully certified. Applicants for this program are taken on a first come, first served basis. You must prove that you are in good physical health, as scuba diving is a rigorous activity. Teacher recommendations are required for first-time applicants. The cost of the summer camp is $1,925. There is no deadline, but camp enrollment is limited to 225 and usually fills up every year. Catalina Island Sea Camp offers two different sessions every summer. To receive a brochure and application for Catalina Island Sea Camp, write, call, or email Guided Discoveries.

- **Guided Discoveries**
  Catalina Island Sea Camp
  PO Box 1360
  Claremont, CA 91711
  Tel: 800-645-1423
  Email: discover@clubnet.com

COLLEGE AND CAREERS PROGRAM AT THE ROCHESTER INSTITUTE OF TECHNOLOGY

College Course/Summer Study

The Rochester Institute of Technology (RIT) gives rising seniors the chance to experience college life and explore a career in the environment by participating in the College and Careers Program, now in its eighth year. Students spend a Friday and Saturday during the summer on campus discussing career options and trying out the institute's facilities. Participants attend four sessions of their choice dealing with a variety of career fields. Of particular interest to those considering the environmental field are such sessions as "Environmental Management: Toxic Spill Response" and a number of offerings in biology, chemistry, and physics. Students are encouraged to stay overnight so as to enjoy the full college experience, but commuters are welcome. Residents pay $45, inclusive of meals and accommodations, while commuters pay $30, inclusive of meals. RIT offers the College and Careers Program twice each summer, usually in mid-July and early August. Applications are due at least one week before the starting date, and the process is competitive. Individual session titles may change from year to year. Contact the Office of Admissions for the latest information and an application form.

College and Careers Program at the Rochester Institute of Technology
Rochester Institute of Technology
Office of Admissions, 60 Lomb Memorial Drive
Rochester, NY 14623-5604
Tel: 716-475-6635

CORNELL UNIVERSITY ENVIRONMENTAL SCIENCES INTERNS PROGRAM

Internship

Cornell University invites students to actively explore the applied sciences in its Environmental Sciences Interns Program. Participants work with Cornell researchers from roughly 9:00 AM to 3:00 PM five days per week on field-based or laboratory research projects. Three days per week, participants join those in Cornell's Exploration in Environmental Sciences Seminar to learn more about careers in the environmental field and how to prepare for them while in high school and college. The Interns Program runs for six weeks in midsummer. Only New York state students who complete their junior year of high school by June of the program year are eligible to apply. Students from ethnic minority groups, with low incomes, from rural areas, or with disabilities are especially encouraged to apply. All applications must be received by the beginning of May. There is no cost for the program and room and board is provided. For more information, contact the Interns Program at Cornell University's Department of Natural Resources.

Cornell University Environmental Sciences Interns Program
Department of Natural Resources, Cornell University
16 Fernow Hall
Ithaca, NY 14853-3001
Tel: 607-255-2807

CORNELL UNIVERSITY SUMMER COLLEGE PROGRAM FOR HIGH SCHOOL STUDENTS

College Course/Summer Study

Cornell University operates a Summer College Program for High School Students that comprises college-level courses and specially designed career exploration seminars. Participants who choose the Exploration in Environmental Sciences Seminar use laboratory and fieldwork to explore the fields of natural resources, ecology, and environmental science. During the program, you are introduced to chemistry and microbiology and learn research, writing, and presentation skills. In addition to the seminar, which is held one afternoon per week, you take two courses, which can include chemistry, geological sciences, microbiology, and natural resources. All courses are college-level and so are far more challenging and demanding than high school courses. Students who have completed their junior or senior year by the June preceding the Summer College Program are eligible to apply for the program, which runs for six weeks from late June to early August. Because of the advanced nature of the course work, Cornell selects participants with proven academic ability, maturity, and intellectual curiosity, as demonstrated by grades, PSAT/SAT/ACT scores, a recommendation from a teacher or counselor, and an application form. Tuition and fees for the Summer College Program total $3,400, and room and board costs an additional $1,600; a limited amount of financial aid is available. Students applying for financial aid should have their applications in by early April; all other applications are due in early May. Cornell University awards full undergraduate credit for the two courses you take but no credit for the career exploration seminar. For more information and an application form, contact the Cornell University Summer College Program.

Cornell University Summer College Program for High School Students
School of Continuing Education and Summer Sessions
B20 Day Hall
Ithaca, NY 14853-2801
Tel: 607-255-6203
Web: http://www.sce.cornell.edu/SC/

Delaware Nature Society

Field Experience, Volunteer Program

The Delaware Nature Society (DNS) offers several volunteer and educational opportunities for high school students. This private, nonprofit organization has been active since 1964 in preserving ecologically important areas of land and water and in educating others about responsible land use and stewardship. The DNS operates from the Ashland Nature Center in Hockessin and also runs the Abbott's Mill Nature Center in Milford. At both locations, volunteers can help maintain nature trails, survey plants and animals, and assist visitors around the grounds and shops. You can also assist at special events such as the Harvest Moon Festival, Native Plant Sale, and Bird Seed Sale. Volunteers can work several days a year or just a few hours at a single event. Those looking to make a stronger commitment might consider volunteering as a Summer Counselor at the Ashland Nature Center. Students ages thirteen to seventeen are eligible to assist teachers as they explain and explore the wonders of nature with children ages three to eight. Volunteers must commit for at least three weeks and be available weekdays from 8 AM to 4 PM. Summer Counselors receive training and enjoy a special end-of-season event, such as a canoe trip or overnight hike.

Young people interested in a lengthier commitment can join the Delaware Stream Watch via the DNS. This involves monitoring a section of stream to determine water quality and chart environmental changes. Stream Watch volunteers must undergo training and monitor their streams at least four times per year, although monthly monitoring is strongly encouraged. For more information on all volunteer opportunities, contact the DNS at the Ashland Nature Center.

Delaware Nature Society
Ashland Nature Center
PO Box 700
Hockessin, DE 19707
Tel: 302-239-2334

EARTHWATCH

Conference, Field Experience, Internship, Membership

Earthwatch is the organization for people whose spirit of adventure is as great as their commitment to the Earth's well-being. A nonprofit membership organization founded in 1972, Earthwatch's major activity is linking volunteers with scientific research expeditions that need them. There are about 130 different expeditions every year, covering all continents but Antarctica, each lasting one to three weeks. At any time from January to October, if you are sixteen or older, you can join an expedition researching Costa Rican caterpillars, for example, or Australia's fossil forests. Whichever expedition you choose, you work with five to ten other people of all ages and backgrounds under the guidance of a research scientist (often a university professor working in his or her field of expertise).

Living and working conditions vary widely among the expeditions; you might stay in a hotel or a tent, remain at one site or hike to several locations while carrying a heavy backpack. Expenses also vary widely, from about $600 to $2,600, depending on travel, accommodation and eating arrangements, and other necessary provisions. Earthwatch reminds potential volunteers, however, that your payment of expenses (along with the donation of your time) is really an investment in environmental research. Of course, you're also investing in your own future. With so many expeditions to choose from, you'll be able to gain experience in career fields from ecology to national park service, from natural history to wildlife preservation. Contact Earthwatch for its annual catalogue listing all the details.

Even if you're not up for one of their demanding expeditions, Earthwatch invites you to become a member at the standard rate of $35 per year. High school students can benefit from the organization's own scholarship, fellowship, and grant opportunities. You can also attend Earthwatch's annual conference, held each autumn in Cambridge, Massachusetts, or apply for an internship at its offices in Boston and San Francisco. Just contact the organization for more information.

Earthwatch
680 Mount Auburn Street, Box 403
Watertown, MA 02272
Tel: 617-926-8200
Fax: 617-926-8532
Web: http://www.earthwatch.org
Email: info@earthwatch.org

ENVIRONMENTAL RESOURCE CENTER

Internship, Volunteer Program

The Environmental Resource Center (ERC), a nonprofit organization founded in 1989, needs volunteers and interns to assist in providing educational programs on environmental issues to the public. The ERC sponsors nine different projects, including Transportation Alternatives for Improved Living, the Waste Reduction and Recycling Initiative, and the Wetlands Education and Restoration Project. Volunteers can assist with any and all of the nine projects as well as with special events and day-to-day operations. You can work as many or as few hours as you like. If you wish to serve as an intern, a more formal time commitment is required, although the specifics are negotiable. The ERC prefers that its interns be at least high school seniors and willing to divide their internship time between a project of their choosing and general office and administrative duties. Prospective interns must submit a resume with references and a description of their interest in the ERC outlining their internship goals. There is the possibility that the ERC may receive grants with which to pay interns a small stipend, but you should expect to work on a voluntary basis. Contact the Executive Director for further information.

Environmental Resource Center
PO Box 819
411 East Sixth Street
Ketchum, ID 83340
Tel: 208-726-4333
Fax: 208-726-1531

ENVIRONMENTAL STUDIES PROGRAM AT THE SCHOOL FOR FIELD STUDIES

Field Experience

The School for Field Studies (SFS) offers summer programs in six locations around the world for students who have completed at least their junior year in high school. You spend one month on location pursuing the four main tenets of the SFS curriculum: an interdisciplinary, "case studies" approach; problem-solving through fieldwork; involvement in the local community; and a proactive role in your own education. You can choose one of the following SFS Centers based on your scientific and environmental interests: Wildlife Management Studies in Kenya, Marine Resource Studies in the British West Indies, Rain Forest Studies in Australia, Sustainable Development Studies in Costa Rica, Coastal Studies on Vancouver Island, and Wetland Studies on Mexico's Pacific Coast. Each program focuses on a genuine environmental problem facing local inhabitants, and your research and suggestions become part of SFS's solution to that problem. All programs are physically demanding and, because there are academic lectures as well as field expeditions, you have virtually no free time. Meals are included in the tuition fee, which averages around $3,000, depending on the location. Extensive financial aid is available. You have to provide some of your own personal backpacking gear and transportation to and from your departure point. The SFS, established in 1981, is fully accredited by Boston University, so it is possible to earn college credit via the summer program. The SFS staff consists largely of trained academics, including many Ph.D.s, with years of field research experience behind them.

The School for Field Studies also offers semester-long programs at all six locations in the fall and spring for students with at least one semester of college credit and one college-level biology or ecology class. All correspondence should be directed to the SFS Admissions Office.

Environmental Studies Program
The School for Field Studies
16 Broadway
Beverly, MA 01915-4499
Tel: 800-989-4418 or 508-927-7777
Fax: 508-927-5127
Email: sfshome@igc.apc.org

Environmental Studies Summer Youth Institute (ESSYI) at Hobart and William Smith Colleges

College Course/Summer Study

Hobart and William Smith Colleges sponsor the Environmental Studies Summer Youth Institute (ESSYI) for rising high school juniors and seniors. Academically talented students are invited to participate in this examination of environmental issues from scientific, social, and humanistic perspectives. Running for two full weeks beginning in late July, the ESSYI is comprised of classroom courses, laboratory procedures, outdoor explorations, and plenty of time to discuss and think about integrating these many approaches to understanding the environment. Lectures encompass ecology, philosophy, geology, literature, topography, and art, among other areas of study, and are conducted by professors from Hobart and William Smith Colleges. Your study of the environment and how humans relate to it also includes field trips to such places as quaking bogs, organic farms, the Adirondack mountains, and Native American historical sites. Participants also make use of the HMS Explorer, the colleges' 65-foot research vessel, as they explore the ecology of nearby Seneca Lake. ESSYI students live on campus and have access to all the colleges' recreational facilities. Those who complete this intellectually and physically challenging program are awarded college credit. For information on costs, financial aid, and application procedures, contact the Institute Director.

Environmental Studies Summer Youth Institute (ESSYI)
Hobart and William Smith Colleges
Geneva, NY 14456-3397
Tel: 315-781-3377
Fax: 315-781-3348

Envirothon

Competition

The Envirothon is a series of competitions, established in 1979, for ninth- through twelfth-graders who want to increase their knowledge of the natural sciences and environmental issues. Progressing from regional to state to national competitions, teams of five students perform experiments and activities and then work together to answer the questions they are given. Competition questions come from five subjects: Wildlife, Forestry, Soil, Aquatics, and Current Environmental Issues. Regional and state competitions feature questions about the local environment, regulations, and concerns in addition to general knowledge questions. While the goal of Envirothon is to develop knowledgeable, environmentally active adults, the program is also designed to be fun for all participants. Teams are headed by high school teachers or other youth leaders and may draw members from schools or from other organizations and associations. At the national level, the Envirothon is sponsored by Canon USA; at the regional levels, usually by state EPAs, forest services, and game and parks commissions. For information about the Envirothon program in your area, contact the relevant state agencies, or speak to your science teacher or guidance counselor. You may also contact the national Executive Director at the address below.

Envirothon
PO Box 855
League City, TX 77574
Tel: 800-825-5547, ext. 27
Email: envirothon@nacdnet.org

FOLSOM CHILDREN'S ZOO AND BOTANICAL GARDENS

Field Experience

The Folsom Children's Zoo and Botanical Gardens offers summer programs in which young people can join either a Junior Zoo Crew (JZC) or an Advanced Zoo Crew (AZC). The JZC program is open to those in seventh grade or above; the AZC is open to those who have participated in the JZC for at least three years. Members of the Junior Zoo Crew spend five full weeks studying and learning how to handle and care for the zoo creatures. The Advanced Zoo Crew program is similar, but participants work more independently and may assume specialized responsibilities as education assistants, shadow veterinarians, or animal care specialists. All JZC and AZC participants learn from and work with Folsom Children's Zoo's staff members and receive written evaluations of their performances. The JZC and AZC programs are nonresidential and are offered twice each summer (one session from June through mid-July, one from mid-July through August). The cost is $65 for zoo members, $70 for nonmembers. Applications are due in mid-February; new applicants to the JZC and all applicants to the AZC must interview as part of the selection process. Contact the Zoo Crew Coordinator at Folsom Children's Zoo and Botanical Gardens for further information or an application.

Folsom Children's Zoo and Botanical Gardens
1222 South 27th Street
Lincoln, NE 68502
Tel: 402-475-6741
Fax: 402-475-6742

FOREST ECOLOGY SUMMER CAMP

Field Study/Summer

The week-long Forest Ecology Summer Camp, run by the University of Minnesota, features a full program of forest stewardship for students ages fourteen to eighteen. Courses include forest management, woodcutting, fisheries, wildlife science, and astronomy. You work in small groups, getting out into the woods to measure trees, take water and soil samples, and learn how to work with a compass. Your mornings are for structured activities, while afternoons are for recreational activities such as supervised canoeing and rock climbing. At the Forest Ecology Summer Camp, you study with the University of Minnesota's natural resource professors, U.S. Forest Service workers, Minnesota's Department of Natural Resources, and local soil and water conservation agencies. Campers sleep four to six in a dorm room and eat in a dining hall. The cost of the program is $220, and a limited number of full scholarships are available. The fee includes everything but transportation to the camp site. Camp enrollment is limited to thirty-five, and the deadline for application is in late June. For registration information, call or write the Forest Ecology Summer Camp.

Forest Ecology Summer Camp
Center for 4-H Youth Development
340 Coffey Hall
1420 Eckles Avenue
St. Paul, MN 55108
Tel: 800-444-4238

FOREST PRODUCTS WEEK ESSAY CONTEST

Competition

All students in Idaho may participate in an essay contest designed to celebrate National Forest Products Week, which takes place in late October. Sponsored by the Idaho Forest Products Commission, the essay contest challenges students—who are grouped by grade—to think about the many different things that forests give us and how we value these things in our lives. Students in the high school group (ninth through twelfth grades) write essays of about five hundred words on topics such as "How Forest Products Touch My Life" and "If I Was in Charge of Our Forests." Students must demonstrate a basic understanding of forest products (they are used by all of us every day, trees are renewable resources, etc.) and must use proper spelling and grammar. The winner in each group receives a $100 savings bond, and his or her class receives $200 to be spent as the winner decides. For current submission dates, essay topics, and entry information, contact the Idaho Forest Products Commission.

Forest Products Week Essay Contest
Idaho Forest Products Commission
350 North Ninth Street, Suite 304
Boise, ID 83702
Tel: 800-ID-WOODS
Fax: 208-334-3449

FRIENDS OF THE EARTH

Membership

This organization is made up of men and women in fifty-two countries working to protect the Earth's well-being. If you're interested in joining their ranks, you can simply pay your dues or go a step further by creating your own local student group. Friends of the Earth is happy to send you a descriptive brochure on starting a student environmental action group. It outlines how to plan and hold your first meeting. It also offers a number of project ideas, such as planting an organic garden and organizing an environmental film series. Once your student group is formed, Friends of the Earth's public information office can provide you with all sorts of useful resources on topics ranging from tackling the waste crisis to preserving marine biodiversity. Most Friends of the Earth initiatives focus on substantive issues with far-reaching effects. For instance, there's currently a cooperative effort with the National Taxpayers Union to cut wasteful and environmentally harmful federal spending. Other projects include a technical assistance program to help citizens influence the operation of city landfills.

Friends of the Earth
1025 Vermont Avenue, NW, Third Floor
Washington, DC 20005-6503
Tel: 202-783-7400
Fax: 202-783-0444
Web: http://www.foe.org/FOE
Email: foe@foe.org

GENESIS FARM

Seminar

Genesis Farm was founded by the Dominican Sisters of Caldwell, New Jersey, in 1980 to provide a site for progressive ecological action and education. The farm comprises 140 acres of woodlands, marshes, and gardens, as well as fields that are organically cultivated by the Sisters on behalf of local families. You can participate in a number of programs and seminars to help you adapt the Sisters' earth-friendly ways to your own gardening and other everyday activities. Genesis Farm's ongoing series of natural foods cooking classes is particularly popular. The spiritual connection between humans and the earth is an important aspect of the Sisters' work and educational projects, and although they are Roman Catholic nuns, they welcome people of all faiths to their programs.

For college students and those interested in continuing education, Genesis Farm offers the Earth Literacy Program. It is a twelve-week module, composed of four short courses, which helps participants "read" what the natural world has to tell us and integrate that experience with intellectual, scientific learning. In conjunction with St. Thomas University in Miami, Florida, Genesis Farm offers the Earth Literacy Certificate, which requires completion of the module as well as an independent ecological study. Undergraduate and graduate students enrolled elsewhere can complete the Earth Literacy twelve-week module, or selected parts of it, for credit through St. Thomas University. For more information on the Earth Literacy Program and college credit possibilities, contact Genesis Farm. For a current calendar of events—including cooking classes and weekend seminars—write, call, or fax Genesis Farm.

Genesis Farm
41 A Silver Lake Road
Blairstown, NJ 07825
Tel: 908-362-6735
Fax: 908-362-9387

GLOBAL RESPONSE

Membership

Global Response is an international action and education network that works with environmental, indigenous, peace, and justice groups. Their mission is to help these groups develop strategies to raise public awareness and create public pressure to address specific environmental emergencies. Every month, Global Response issues an Action bulletin to its members. Eco-Club Actions is the monthly bulletin specifically for high school students and youth environmental clubs. It describes an environmental emergency, gives background information, recommends points to be made in a letter, and provides the names, addresses, fax numbers, and email addresses of people who are in a position to make positive changes. Then it's up to you to write your own letter requesting that action be taken. The idea is for small groups of students, perhaps getting together to share a pizza, to sit down and write these letters. This simple approach is highly effective. Global Response played a major role in creating a wetlands park in South Africa and stopping the dredge mining within it. They also saved nine hundred thousand acres on Honduras's Mosquito Coast from clearcutting by a lumber company. Contact Global Response for more information and the latest Eco-Club Actions newsletter. If you are interested in starting your own environmental club, request The Eco-Club Activist Guide.

Global Response
PO Box 7490
Boulder, CO 80306-7490
Tel: 303-444-0306
Web: http://www.globalresponse.org
Email: globresponse@igc.apc.org

GREENPEACE USA

Internship, Membership

Greenpeace offers general memberships as well as internships to those who are interested in joining its high-profile campaign of environmental activism. You and professional Greenpeace staff decide jointly how long the internship will last, when it will take place, and what duties it will entail. However, all interns receive a formal orientation to Greenpeace, and must read the book, *The Greenpeace Story,* and view several videos as part of their introduction to the organization. As an intern, you may work in all five divisions of Greenpeace (Campaigns, Development/Media, Public Outreach, Finance, and Administration) or choose to specialize in just one or two of them, depending on your personal interests. You must remain flexible to switch to projects where you are most needed, and should be committed to attending outreach activities such as demonstrations and distributing informational handouts. All internships are on a strictly volunteer basis. To be considered for an internship, you must submit a resume, three letters of reference, and a letter detailing your interest in a Greenpeace internship. Because each internship is individualized, there is no general application deadline. If you are interested in pursuing an internship with Greenpeace, contact its Director of Human Resources. The office staff will handle general requests for membership information.

Greenpeace USA
1436 U Street, NW
Washington, DC 20009
Tel: 202-462-1177
Fax: 202-462-4507

GREEN TEENS PROGRAM

Volunteer Program

Green Teens matches young people between the ages of sixteen and eighteen with volunteer service opportunities in one of Ohio's seventy-two state parks.

Possible projects include environmental cleanup, prairie plantings, and constructing wildlife feeding stations. The parks offer prepackaged projects but are also happy to consider your own ideas. Volunteers can participate as interested individuals or as part of a group (nature club, ecology class, etc.). You can work as little as one hour or as long as one year, and projects can be arranged for weekdays, weekends, or evenings any time of year. A volunteer contract agreement specifying guidelines for the project and its duration must be signed by the volunteers and the park manager. You may have to provide your own tools and/or materials, depending on the project. Besides conservation experience, Green Teens receive worker's compensation coverage, newsletters, and field trip opportunities. The program, up and running since 1993, is operated by the Ohio Department of Natural Resources, Division of Parks and Recreation, and staffed by associates of the Ohio State Parks. Contact either the ODNR Volunteer Coordinator or your local state park for more information.

Green Teens Program
Division of Parks and Recreation
Ohio Department of Natural Resources
1952 Belcher Drive, Building C-3
Columbus, OH 43224-1386
Tel: 614-265-6549
Fax: 614-261-8407

High School Field Ecology Program at the Teton Science School

College Course/Summer Study, Field Experience

The Teton Science School operates an intense High School Field Ecology summer program for students ages sixteen to eighteen. The program lasts six weeks and, while based in Jackson Hole, WY, includes time at Yellowstone and Grand Teton National Parks. In the program, you "learn field ecology by doing field ecology," and are expected to maintain a high level of physical activity while conducting research and exploring the ecosystem. Working with instructors, scientists, and representatives of federal agencies, you learn proper field investigation techniques as well as how to keep a field journal, read and make maps, and explore the wilderness without harming the environment. The program culminates with a major group research project that is written up and presented to a group of scientists and other guests. Each participant is formally assigned a grade and given a written evaluation for his or her work in the program, so this is a suitable source of high school credit. You live in log cabin-style dormitories with modern facilities, except during a week-long backpacking trip through the mountains. Most outdoor gear and all meals are covered by the tuition fee of around $2,500; some financial assistance is available. This program accepts only thirteen to fifteen people each year, and you must have completed biology and received favorable teacher recommendations to be considered. Applications are generally accepted between January 1 and March 15. The Teton Science School also runs a Junior High School Field Ecology program, a Field Natural History course for fourteen- to sixteen-year-olds, and various programs for adults and teachers. All correspondence should be directed to the Registrar.

High School Field Ecology Program
Teton Science School
PO Box 68
Kelly, WY 83011
Tel: 307-733-4765
Fax: 307-739-9388

ILLINOIS DEPARTMENT OF NATURAL RESOURCES

Volunteer Program

The Illinois Department of Natural Resources (IDNR) runs a Volunteer Network open to high school students interested in protecting the state's natural resources and helping visitors to appreciate them and enjoy them responsibly. Because the Network encompasses all state parks, fish and wildlife areas, conservation areas, and state forests, there are opportunities available nearby regardless of where you live in Illinois. Volunteers can also assist with various programs and projects run by the IDNR. The Illinois Department of Natural Resources offers internships, some paid, for college students. The Volunteer Network welcomes participants of all ages. For more detailed information on available positions, consult the Volunteer Project Directory, available from the Illinois Department of Natural Resources, or call the IDNR directly. If you do not live in Illinois, contact your own state's department of natural resources; they are likely to offer similar opportunities.

Illinois Department of Natural Resources
524 South Second Street
Springfield, IL 62701-1787
Tel: 217-785-0067

INSTITUTE FOR EARTH EDUCATION

Membership

The Institute for Earth Education (IEE) is committed to improving the environmental problems of the world through an educational approach. At the heart of IEE is an emphasis on developing deep relationships with the Earth and on achieving tangible outcomes. Members receive the seasonal journal, *Talking Leaves,* which addresses topics that spark debate and discussion. IEE has special student clubs that span all age ranges. Sunship III is the newest group, and it's specifically for high school students. As this club is just getting started, no specifics were available at the time this book went to press. Sunship III's goals and projects should soon be available on IEE's Web site, or you can write or call the organization directly for more information.

Institute for Earth Education
Cedar Cove
Greenville, WV 24945
Tel: 304-832-6404
Web: http://www.slnet.com/cip/iee
Email: iee1@aol.com

The Jane Goodall Institute—USA

Membership

In 1960, anthropologist Jane Goodall ventured into East Africa to begin a field study on wild chimpanzees—and she's still at it today. Her ongoing research constitutes the longest continual field study of animals in their natural habitat. The Jane Goodall Institute was founded in 1977 to work on wildlife research, environmental education, and the conservation and welfare of animals, especially chimpanzees. The Institute's Roots & Shoots club is for students from preschool to university age. Its mission encompasses both humanitarian and environmental issues. When you sign up to be a member, you join other students in thirty-eight U.S. states and more than thirty countries. These groups are extremely active. In 1997, they organized International Roots & Shoots Peace Day, with events and projects held in local communities to promote respect, consideration, and understanding among people. As a member, you'll receive the semiannual Roots & Shoots Network publication and the annual *Jane Goodall Institute World Report.* Dues are $25.00 per year for Roots & Shoots.

The Jane Goodall Institute—USA
PO Box 599
Ridgefield, CT 06877
Tel: 203-431-2099
Web: http://www.wcsu.ctstateu.edu/cyberchimp/
Email: janegoodall@wcsu.ctstateu.edu

LOUISIANA NATURE CENTER

Volunteer Program

The Louisiana Nature Center, founded in 1981, is a nonprofit educational center affiliated with the Audubon Institute. The Center, located on wetlands within the Joe W. Brown Memorial Park, runs an Apprentice Volunteer Program year-round for young adults who want to learn about the environment and share their knowledge with others. Participants must be between twelve and seventeen years of age, have a GPA of at least 3.0 and two recommendations, as well as a positive attitude and a genuine appreciation of nature. Once accepted into the program, you must attend a general orientation class and then undergo specialized training in the department to which you are assigned. You might care for animals, interpret the setup at the nature station for visitors, help others make nature crafts, or assist in special wetlands maintenance activities, among other possibilities. You need work only one four-hour shift every other Saturday or Sunday, so this is a good opportunity even if you are very busy with other activities. Besides learning about wetlands wildlife and plants and working with environmental professionals, other benefits of the Apprentice Volunteer Program include free admission to the Louisiana Nature Center and free or reduced admission to other Audubon Institute agencies.

The Louisiana Nature Center also welcomes adults ages eighteen and older into its general volunteer program. Many positions are available, including animal care assistant, grounds worker, librarian, and naturalist, and all the necessary training is provided. The Center also offers a number of programs for younger children, particularly student groups. Interested teachers or parents should write or call for a copy of the Louisiana Nature Center's current Educational Services Guide. For further information, contact the Volunteer Coordinator at the Louisiana Nature Center.

Louisiana Nature Center
PO Box 870610
New Orleans, LA 70187-0610
Tel: 504-246-5672
Fax: 504-242-1889

NATIONAL WILDLIFE FEDERATION TEEN ADVENTURE PROGRAM

Camp, Field Experience, Membership

The National Wildlife Federation operates a Teen Adventure program for those between fourteen and seventeen years of age. One- to two-week sessions are held throughout the summer at Hendersonville, North Carolina,and Estes Park, Colorado. Each session focuses on environmental education and stewardship, outdoor living skills, service projects, and teamwork. The specific activities and wildlife studies vary by location, so you can concentrate on the environment of either the Blue Ridge Mountains or the Rockies. Students who have already completed one Teen Adventure may attend one of the special Superstar sessions in North Carolina to learn advanced skills and participate in a special Career Day on wildlife conservation. All programs are limited to between eight and twelve people per session and are led by the trained staff of the National Wildlife Federation. You must supply some of your own equipment, including a sleeping bag, backpack, and hiking boots. Tuition, including all meals, ranges from $620 to $850 depending on the length and location of the session. Early-bird pricing is available to those who apply before March 1.

There are two levels of student memberships. The basic membership is $9 per year, which gets you *EnviroAction,* NWF's environmental news digest, a personalized membership card, a member decal, and every edition of NWF's Wildlife Stamps. For $12 a year, the associate membership includes a six-issue subscription to either *National Wildlife* or *International Wildlife* magazine.

The National Wildlife Federation also offers special programs for junior high school students and Conservation Summits for entire families. Of course, membership is open to all adults as well. Inquiries and applications for the Teen Adventure Program should be directed to the Registrar. If you want to join the NWF, contact the Membership Office.

National Wildlife Federation Teen Adventure Program
8925 Leesburg Pike
Vienna, VA 22184-0001
Tel: 703-790-4000
Web: http://www.nwf.org/

NATURAL RESOURCES CAREER WORKSHOP

College Course/Summer Study, Field Experience

The Central Wisconsin Environmental Station, in cooperation with the University of Wisconsin-Stevens Point (UW-SP), offers a Natural Resources Careers Workshop for high school students who have completed at least their sophomore year. During this five-day residential program, you explore careers in forestry, wildlife management, soil and water science, park management, environmental education, and other areas. Natural resource professionals, including UW-SP faculty and specialists from federal resource management agencies, lead presentations and discussions on the many career opportunities available. To experience the work firsthand, you will participate in excursions to the UW-SP College of Natural Resources, a water treatment plant, a recycling center, and various parks. Other activities include water and soil sampling, wildlife identification, canoeing, swimming, and hiking. The workshop is held four times each summer, with one special session offering "A Cross Cultural Perspective" on natural resources careers from people of many different backgrounds. The program costs roughly $250, including all room, board, and program fees. Each session is limited to fifty participants.

The Central Wisconsin Environmental Station offers a number of programs for younger children and adults during the summer, and holds family activities throughout the year.

Natural Resources Career Workshop
Central Wisconsin Environmental Station
7290 County MM
Amherst Junction, WI 54407
Tel: 715-824-2428
Fax: 715-824-3201

NEW JERSEY GOVERNOR'S SCHOOL ON THE ENVIRONMENT

College Course/Summer Study

The New Jersey Governor's School on the Environment is a summer program for gifted and talented rising seniors. To participate in this program, you do not necessarily have to have an interest in scientific matters as the program explores environmental issues from social, economic, and political as well as scientific perspectives. Guided by experienced high school and college faculty, you perform research activities on such topics as pollution and public health, pesticides, environmental law, and global ecosystems. You attend daily forums to exchange ideas with other students and faculty, and evening lectures and presentations on state and local issues. There are also opportunities to watch and participate in performing arts events, and you experience campus life by living at Stockton State College. Admission to the Governor's School on the Environment is highly selective. You must first be recommended by your principal and/or guidance counselor, who consider GPA, test scores, curriculum, and extra-curricular activities to determine whether you possess the established characteristics of the talented and gifted. Their recommendations along with essays you have written are then considered by county and state selection committees, which ultimately decide who will attend. This residential program is fully funded by the New Jersey Department of Education. In New Jersey, contact the Governor's School on the Environment or your principal or guidance counselor for further details. Outside New Jersey, contact your governor's office or state Department of Education to see if a similar program is available.

New Jersey Governor's School on the Environment
Stockton State College
Pomona, NJ 08240-9988
Tel: 609-652-4924

Oceanography Summer Program at Occidental College

College Course/Summer Study

Occidental College offers a summer program in oceanology, begun in 1975, using the staff and facilities of the college's Laboratory of Zoology and Department of Biology. This four-week, residential program is open to rising seniors who have completed courses in mathematics, biology, and chemistry or physics. During the program, you participate in lectures, labs, and fieldwork on chemical and physical oceanography, classification and natural history of many marine plants and creatures, and marine ecology and the effects of pollution. You also go on field trips, camping trips, and short cruises, and enjoy social trips to the beach and to Disneyland. All participants live on the Occidental campus, where they can discuss how to choose a college and how to apply to colleges with Occidental's admissions staff. Only twenty-four students are accepted each summer; there is no formal deadline, but applications are evaluated on a first come, first served basis. Tuition costs about $2,700; room, board, and other miscellaneous expenses total an additional $1,000. Students who complete the program receive full credit from Occidental College. Contact the Program Director, at the Department of Biology, for further details.

Oceanography Summer Program
Occidental College
Biology Department, 1600 Campus Road
Los Angeles, CA 90041-3314
Tel: 213-259-2890
Fax: 213-341-4974

Oregon Museum of Science and Industry

Camp, College Course/Summer Study, Field Experience

The Oregon Museum of Science and Industry (OMSI) offers many study and adventure opportunities both at its riverfront facility in Portland and at field stations and remote locations in the Pacific Northwest. You can take one- or two-hour classes at the museum in Portland throughout the year; recent classes have included "Inside Insects" and "Animal Innards." Hour-long classes cost $100 while those lasting two hours cost $180. All high school students are eligible for these classes. During the summer, OMSI offers a dozen different Junior Naturalist Camps around Oregon and Washington, including a backpack adventure, a raft adventure, and the Natural Science Field Study and Research Program. Each camp lasts one to two weeks and is open to all those between fourteen and eighteen years of age, although some programs require advanced wilderness skills and considerable physical stamina. You must provide your own camping equipment and, on some advanced programs, your own food and cooking skills. It may be possible to earn high school credit for completing a Junior Naturalist Camp. The average fee for these adventures is around $300, although there is a discount for students whose families are members of OMSI, and some financial assistance is available. All programs are closely supervised by OMSI staff, college-educated scientists who are also certified in CPR and first aid.

OMSI also offers classes and learning adventures for children as young as five and for adults. Teachers can arrange special projects at the Museum (including overnight stays for their students) and visits by OMSI staff to their schools. Direct correspondence to the OMSI Camp Registrar.

Oregon Museum of Science and Industry (OMSI)
1945 SE Water Avenue
Portland, OR 97214-3354
Tel: 888-674-OMSI or 503-797-4547
Fax: 503-239-7800
Web: http://www.omsi.edu/

Peregrine Fund at the World Center for Birds of Prey

Volunteer Program

The Peregrine Fund has worked nationally and internationally to promote conservation and environmental education since 1970, "focusing on birds to conserve nature." The Fund is headquartered at the World Center for Birds of Prey, which houses over two hundred birds and releases their young into the wild. The Velma Morrison Interpretive Center at the headquarters provides opportunities for the public to learn more about birds of prey by encountering live birds, multimedia presentations and exhibits. Volunteers are needed to make presentations and lead tours of the Interpretive Center. To qualify, you must be at least sixteen years old, a resident of the area, and able to devote a minimum of four hours per week to the center. Additionally, to handle the birds and serve as a docent, you must participate in roughly twenty-five hours of training, so a genuine commitment to the work is vital. Those who would like to volunteer but are available only for a shorter training period may find opportunities in the gift shop, the reception area, or the office. Continuing education programs are offered to all volunteers who want to learn more about birds of prey and general conservation work. Training classes are scheduled annually, but applications are accepted year-round and individual training is also available.

The Velma Morrison Interpretive Center accepts volunteers of any age over sixteen provided they are willing to participate in the necessary classes and training exercises. For more information about volunteer opportunities, contact the Director of Volunteers in care of The Peregrine Fund.

Peregrine Fund at the World Center for Birds of Prey
Velma Morrison Interpretive Center
566 West Flying Hawk Lane
Boise, ID 83709
Tel: 208-362-8687
Fax: 208-362-2376
Web: http://www.peregrinefund.org
Email: tpf@peregrinefund.org

PLANET DRUM FOUNDATION

Internship, Volunteer Program

The Planet Drum Foundation was founded in 1973 to promote the ecological concept of bioregions: areas of land grouped according to shared geological and climatological characteristics, rather than arbitrary political boundaries. It is ostensibly easier to plot environmental policy and determine sustainability when using the natural divisions of bioregions. Planet Drum welcomes volunteers and interns to work in spreading the bioregion concept and supporting their Green City Project, which matches concerned citizens with local environmental groups in need of volunteers. During a Planet Drum internship, you can work as an assistant in Administration, Art and Design, Editing, Fundraising, Outreach/Membership, or other departments of the organization. Necessary qualifications vary with the position, but the main requirements are dedication to environmental concerns and a willingness to learn and work. Planet Drum prefers that its interns—who are unpaid—make a three-month time commitment of twenty hours per week. The Foundation gladly provides letters of recommendation at the end of your internship, providing your work warrants them. Those who cannot make such a time commitment can still work at Planet Drum as a volunteer, on a regular schedule or whenever they are most needed. Interns who do not live in the San Francisco Bay area cannot get housing through the Foundation, but they will try to help you find it.

The Planet Drum Foundation/Green City Project welcomes adult volunteers of all ages. It is happy to help college students who complete internships with them receive credit for their work. Contact the Volunteer

Coordinator for more information on Planet Drum and its volunteer opportunities.

Planet Drum Foundation
Box 31251
San Francisco, CA 94131
Tel: 415-285-6556
Fax: 415-285-6563

Project ORCA/Marine Adventure Camp

Camp, College Course/Summer Study, Field Experience

The Island Institute, founded in 1989, runs Project ORCA for all teenagers who are interested in combining marine science study with a camping adventure. While participating in this ten-day residential program, you "adopt" a marine animal native to the San Juan Archipelago, just off the coast of Washington state. You learn scientific methodologies while tracking and researching your chosen animal. Your self-designed program may also include visits to local research centers and expeditions at sea and on shore. All participants enjoy whale watching, snorkeling, sea kayaking, hiking, and excursions to the various San Juan Islands. You live in tenting accommodations and take meals at a central lodge on Stuart Island. Accommodation, food, and equipment are all included in the tuition fee of around $1,100. Project ORCA operates only twice per summer, and space is limited to fifteen participants per session.

The Island Institute also operates a seven-day ORCA Institute for adults over eighteen years of age, and Family Camps for children and adults of all ages. Contact the Island Institute for further information and specific dates.

Project ORCA/Marine Adventure Camp
Island Institute
PO Box 1099
Friday Harbor, WA 98250
Tel: 800-956-ORCA
Fax: 206-463-3396
Email: isleinst@aol.com

Rainforest Action Network

Membership

The Rainforest Action Network (RAN) is a 30,000-member grassroots organization dedicated to stopping the destruction of tropical rain forests. You can become involved with RAN on an individual basis by simply checking their Web site frequently and responding to "Action Alerts." Or you can join forces with other activists in a Rainforest Action Group. RAGs, as they're called, are community or student groups of concerned people. They work through direct action, letter writing, protests, and by pressuring corporations and politicians. There are more than 125 RAGs; in fact, one of the most active groups is headed by a high school student. If you decide to start your own group, RAN will provide information, advice, event notification, ideas for activities, a quarterly report, and a bimonthly newsletter. Dues are $35 per year. Visit the Rainforest Action Network on the Web or write for more information.

Rainforest Action Network
221 Pine Street, #500
San Francisco, CA 94104
Tel: 415-398-4404
Web: http://www.ran.org
Email: ran-info@ran.org

SEACAMP SAN DIEGO

Camp, College Course/Summer Study

Seacamp, founded in 1987, offers summer camps for those ages twelve to eighteen, using the Pacific Ocean as a classroom for marine science education. Campuses in San Diego and Hawaii are fully equipped with modern residential facilities as well as an aquarium room and laboratory. Staff members hold degrees in marine science and are trained in safety and first aid. Seacamp is offered as a residential or day program and includes labs and workshops in marine biology and ecology, oceanography, and a study of career opportunities; field trips, tidepooling, snorkeling, and an introduction to scuba diving round out the curriculum. Seacamp II is a residential program for those who have completed Seacamp and want to tackle a more in-depth marine science research project and improve their diving and snorkeling capabilities. The week-long camps are offered several times each summer at both locations. The number of places available in all programs is limited. The tuition fees are all-inclusive and average about $500 for the Seacamp day program, $650 for residential, $750 for Seacamp II in San Diego, and $950 for Seacamp II in Hawaii. There is a special offer for students who take Seacamp and Seacamp II back-to-back.

Seacamp also offers programs for visiting classes during the school year, traveling programs for area schools, and programs for adults.

Seacamp San Diego
Seacamp Enterprises, Inc.
3669 Mount Ariane Drive
San Diego, CA 92111-3904
Tel: 800-SEACAMP
Fax: 619-569-7011
Web: http://www.seacamp.com/
Email: admission@sea.edu

SEA WORLD/BUSCH GARDENS ADVENTURE CAMPS

Camp

Sea World and Busch Gardens together offer a number of adventure camps for high school students each summer. For those who are seriously considering an environmental career, Sea World of Texas runs Careers Camps and Advanced Careers Camps from March through October. These week-long programs allow participants to go behind the scenes and assist animal care professionals as they work with Sea World's own residents. You are instructed in the work by the professionals and have the opportunity to discuss this field with these and other experts. Animal protection and conservation issues are among Sea World's priorities, so they play a large part in this camp. Careers Camps are open to those presently in the ninth to twelfth grades, who must submit an application, transcript, and recommendation and participate in an interview over the telephone. Advanced Careers Camps are for current tenth- through twelfth-graders who have already completed a Careers Camp; other application requirements are the same. These are fun, hands-on programs that also take academic and career education seriously. Some high schools award credit for completing the camps, so check with your guidance counselor. Camp fees start around $650 but are usually higher during the peak summer season. Fees include tuition, room and board, supplies and equipment, and transportation to and from San Antonio International Airport. Airfare is extra, but Southwest Airlines offers substantial discounts.

Sea World in California and Sea World/Busch Gardens in Florida offer other camps related to the environment but without a career focus. Contact them directly for further information. Contact Sea World of Texas for information on the Careers Camps and Advanced Careers Camps.

Sea World/Busch Gardens Adventure Camp
Sea World of Texas
10500 Sea World Drive
San Antonio, TX 78251-3002
Tel: 800-700-7786 or 210-523-3608
Web: http://www.seaworld.org

SIERRA STUDENT COALITION

College Course/Summer Study, Membership

The Sierra Student Coalition (SSC) was founded by a high school student in 1991 as the student-run arm of the Sierra Club, America's oldest and largest environmental organization. High school students are welcome to join the SSC as individuals or as part of a group (such as an ecology club or scouting troop). Student dues are $15 a year. The SSC develops grassroots campaigns on environmental issues, working largely through a network of activists and coordinators who organize campaigns at all levels. As a member, you receive *Generation E,* a newsletter full of action opportunities, organizing advice, and articles on student activism. Each semester, the SSC runs major national campaigns and local projects that you can work on or even lead. The SSC is also a great resource if you're already part of an existing club. The newsletter should give you great ideas and boost your enthusiasm for ongoing projects. Members can also link up with the Sierra Club's speakers and get help in capturing media attention for local campaigns.

All SSC members entering the ninth through twelfth grades may take part in its summer High School Environmental Leadership Training Program. The program is held in early August in Vermont and in California (there are two separate sessions so that participants can attend a gathering relatively close to them, but the content is identical). Each session runs for six days and five nights, during which experienced student activists help you learn how and when to take action and how to recruit others to support your cause. Each session includes seminars on specific topics such as Effective Public Speaking, Creative Fund-Raising, and How to Make a Group More Effective. Because both sessions of the Leadership Training Program are held in rustic settings with woodsy or rural surroundings, a good deal of time is also devoted to the exploration and appreciation of the natural world. Room and board, entertainment expenses, and local transportation to and from the site are all included in the program fee of $135 (nonmembers must join the SSC and pay the $15 membership fee, too). Some scholarships are available for those who otherwise could not attend due to financial need. Applications are accepted on a rolling basis through the end of April, but because space is limited you are encouraged to apply earlier. You may apply to both sessions to increase your chances of securing a place, but you may attend only one. For further information and an application for the High School Environmental Leadership Training Program, contact the SSC Summer Program.

Sierra Student Coalition
PO Box 2402
Providence, RI 02906
Tel: 401-861-6012
Fax: 401-861-6241
Web: http://www.ssc.org/
Email: chapter@spusa.org

SUMMER PROGRAM IN MARINE STUDIES AT THE ACADIA INSTITUTE OF OCEANOGRAPHY

College Course/Summer Study

The Acadia Institute of Oceanography, offering educational programming since 1975, runs two-week residential marine science programs every summer. Basic Sessions are offered three times per year for twelve- to fifteen-year-olds. In these sessions, you will take a solid natural history approach to oceanography, learning basic marine concepts and becoming acquainted with the methods and materials of laboratory research. Advanced Sessions are offered twice per year for fifteen- to eighteen-year-olds who have completed at least high school biology or chemistry. Students in these sessions must prepare regular laboratory reports and analyses of their work. You are also introduced to professionals currently working in the field and have the opportunity to explore careers in marine science. Students use the institute's own wet-laboratory, survey vessel, and other facilities throughout the program. All sessions are primarily academic (although there is some recreation time each day), but no one is expected to have any previous training in oceanography. Students live at the Acadia Institute on Mount Desert Island in Seal Harbor, Maine. Transportation to and from the mainland, room, board, and virtually all other expenses are included in the tuition fee of about $1,300. Exemplary students who complete the Advanced Session may be eligible for a special winter course in marine science. The Acadia Institute for Oceanography will give further details to those it deems suitable.

Summer Program in Marine Studies at the Acadia Institute of Oceanography
Acadia Institute of Oceanography
PO Box 98
South Berwick, ME 03908-0098
Tel: 207-384-4155
Fax: same
Email: AI01@aol.com

SUMMER WORKSHOPS FOR HIGH SCHOOL STUDENTS AT SOUTHAMPTON COLLEGE

College Course/Summer Study

Southampton College of Long Island University offers residential summer programs in marine science and environmental science for rising juniors and seniors. The programs are taught by Southampton College professors, and counselors provide twenty-four-hour supervision. The Introduction to Marine Science program offers field experience in marine ecology, coastal geology, physical oceanography, and ichthyology. Participants are taught proper measurement and sampling methods, and use their knowledge during excursions in the college's own research vessels. If you are interested in a more generalized overview of terrestrial and marine environments you may prefer Long Island Environment: Resources and Issues, Problems and Solutions. In this program, you explore local ecosystems, the threats they face, and possible methods of safeguarding or restoring them. Both programs last one week and are offered in late July and August. The cost is $300, which includes room, board, tuition, and supplies; some evening activities may cost extra. Students sleep two per room in dormitories. To apply, you must submit a completed application form, high school transcript, your PSAT/SAT scores, and a writing sample by late May. Contact the Summer Office for a form or further details.

Summer Workshops for High School Students
Long Island University, Southampton College
239 Montauk Highway
Southampton, NY 11968-4198
Tel: 516-287-8349
Web: http://www.southampton.liunet.edu
Email: info@southampton.liunet.com

TREE MUSKETEERS

Membership, Volunteer Program

Back in 1987, a group of young Girl Scouts felt so bad about using paper plates on a camping trip, they decided to plant a tree to make up for it. That was the beginning of the Tree Musketeers, a group whose mission is to empower young people to lead environmental improvement. All its activities are led by and for young people with adults serving as a support system. Tree Musketeers offers a range of volunteer opportunities for interested young people. Volunteers are needed to assist at tree plantings and tree care parties, organize regional and national summits, and take part in special projects such as mass mailings, neighborhood canvassing, collecting seeds, and writing for the newsletter. You can also help out with office work, including answering telephone and mail inquiries, data entry, and project management. Other volunteer opportunities often arise with little advance notice, so you can sign on to the volunteer notification list to keep posted on the latest developments. If you are looking for a formal volunteer commitment, the Leadership Empowerment & Action Development (LEAD) program provides the basis for a three-month environmental project. Small groups of young people, perhaps already organized as a scouting troop or eco-club, work with adult partners on a project of their own creation or on one that is connected to an ongoing Tree Musketeers program.

Currently there are two major initiatives for you to join: the Hometown Program and the Partners for the Planet Network. The Hometown Program encourages young people to lead urban forestry projects in their own hometowns. Call for an Action Kit, with all sorts of information on tree planting and care, composting, and starting your own group. Partners for the Planet is a large network that offers services to students who are already involved in environmental clubs. For instance, you could sign up to attend a National or Regional Youth Summit, where kids gather to learn from one another. Another service—the Youth Hotline—gives you toll-free access to excellent environmental resources. For more information about Tree Musketeers, its volunteer opportunities, and the LEAD program, contact the main office.

Tree Musketeers
136 Main Street
El Segundo, CA 90245
Tel: 310-322-0263

U.S. Forest Service

Employment Opportunity, Field Experience

The U.S.D.A. Forest Service is the government's major conservation organization and the world's largest forestry research group; it manages public lands in national forests and grasslands on 191 million acres. The Forest Service runs the Student Career Experience Program (SCEP) for people at least sixteen years old who are enrolled in an accredited high school, vocational school, or college. SCEP offers its participants the chance to gain paid work experience in forestry stewardship while completing school. In the program, you work part- or full-time for a total of at least 640 hours in your local national forest or grassland. Pay generally ranges from $7 to $11 per hour, and participants are usually eligible for vacation time, sick leave, retirement benefits, life and health insurance, and reimbursement of some expenses. If the program is successfully completed before graduation, the Forest Service generally makes an offer of permanent employment; however, many positions are restricted to those who are graduating from college with a bachelor's degree. The Student Temporary

Employment Program (STEP) is very similar to SCEP except that it does not lead to permanent employment, carries no insurance or retirement benefits, and does not require college students to work in positions related to their majors. Students who begin with STEP can transfer into SCEP and apply their work experience to the 640-hour total. For further information, contact your school's career planning and placement office or your local Forest Service employment office.

The U.S.D.A. Forest Service also relies on volunteers to help care for the land and to work with the many visitors who arrive each year. Contact your nearest Forest Service Office to learn more about volunteer opportunities, or contact the national office listed below.

U.S. Forest Service
U.S. Department of Agriculture
PO Box 96090
Washington, DC 20090-6090
Web: http://www.fs.fed.us

U.S. Forest Service at San Juan-Rio Grande National Forests

Internship, Volunteer Program

The U.S. Forest Service at San Juan-Rio Grande National Forests has many volunteer positions and internships available. A booklet called "Volunteer Opportunities," issued at the beginning of every calendar year, lists expected openings for the coming twelve months. Positions range from Animal Surveyor to Soil Science Assistant to Campground Interpreter. You can, however, create your own position by working with the Volunteer Coordinator; help is needed in such areas as botany, fish/wildlife, natural resources planning, and timber/fire prevention. Qualifications vary depending on the position desired. Volunteers and interns are needed year-round, and projects can last from a few days to a few months. Such amenities as housing, food, transportation to site, and personal gear may or may not be provided, depending on the project. This program has been growing steadily since its inception five years ago and now can sometimes accommodate out-of-state volunteers.

San Juan-Rio Grande National Forests have volunteer and intern positions available for adults of all ages. Applications are accepted year-round and should be sent to the Volunteer Coordinator.

U.S. Forest Service at San Juan-Rio Grande National Forests
1803 West Highway 160
Monte Vista, CO 81144
Tel: 719-852-5941
TTY: 719-852-6271
Fax: 719-852-6250

WILDLIFE ENVIRONMENTAL SCIENCE INTERNSHIP PROGRAM

Internship, Volunteer Program

The Chesapeake Wildlife Sanctuary, a nonprofit wildlife rehabilitation center open 365 days per year, offers internships to high school students who want to assist in its work. As an intern, your duties might include answering the wildlife rescue hotline, cleaning cages, feeding and caring for wildlife, and preparing research reports on wildlife care and rehabilitation. It is possible to gain high school credit through one of these internships. Applications are accepted year-round, but most internships begin at the start of a semester or the summer. It is possible—and advisable—to apply as much as one year in advance of your expected starting date. You must be available at least sixteen hours (eight hours two days per week or four hours four days per week) on the same days of the week every week for at least three months. Such consistent scheduling is crucial to the proper care of the animals. Internships are strictly on a voluntary basis, although students who work full-time (forty hours per week for twelve weeks) can apply for a merit-based scholarship of $200 to $350 at the conclusion of their internship. All students must supply their own transportation to and from the Chesapeake Wildlife Sanctuary.

The Chesapeake Wildlife Sanctuary offers more demanding internships (with credit possible) for college students and has volunteer opportunities for adults of all ages. Further information about internships and less formal volunteer positions is available from the Sanctuary.

Wildlife Environmental Science Internship Program
Chesapeake Wildlife Sanctuary
17308 Queen Anne Bridge Road
Bowie, MD 20716-9053
Tel: 301-390-7010
Email: cheswild@erols.com

YOSEMITE BACKCOUNTRY ADVENTURE

Camp, Field Experience

The Yosemite Institute, established in 1971, works in cooperation with the National Park Service to offer the Yosemite Backcountry Adventure to students between the ages of fourteen and seventeen. This is a two-week residential course offered twice every summer. Led by professional naturalist guides, you explore Yosemite's high peaks, deep canyons, alpine lakes, and other features rarely seen by other visitors. You learn about the area's abundant wildlife and unique cultural and natural history while hiking four to six miles per day at elevations of six to ten thousand feet. Only twelve participants are accepted for each Yosemite Backcountry Adventure. The cost is roughly $1,000 per student, including meals and group overnight gear (tents, cooking pots, etc.). You must provide your own personal gear, however, including sleeping bag, water bottle, and utensils.

The Yosemite Institute also offers environmental workshops for teachers, and various programs throughout the year. Contact the institute for further information and for details on available scholarship funds.

Yosemite Backcountry Adventure
Yosemite Institute
Box 487
Yosemite, CA 95389
Tel: 209-379-9511
Fax: 209-379-9510

YOUTH FOR ENVIRONMENTAL SANITY! (YES!)

Camp, Membership

Youth for Environmental Sanity! (YES!) is a nonprofit project run by youth between the ages of seventeen and twenty-five. Its mission is to educate, inspire, and empower young people to take positive action for the future of life on earth. YES! has managed summer camps for young people between the ages of fifteen and twenty-two since 1991. There are now six different week-long camps at various locations around the country, with each camp's character determined by its location. The camp at Stillwater, Oklahoma, is held on lands sacred to Native Americans; campers at Nevada City, California, are surrounded by the Tahoe National Forest; the Friday Harbor Island camp takes place off the Washington coast; participants are just two hours away from downtown Los Angeles at Camp Osceola in Angelus Oaks, California; Sun Point Farm in Derry, New Hampshire, makes a charming, rustic campsite; and Whitefish, Montana, is the home of the popular African Rhythm Camp. While the program varies by location, all camps feature guest presenters with extensive experience in environmental activism, along with training in fundraising, public speaking, and obtaining media coverage, and learning how to build lasting communities and friendships. Participants are asked to pay for the costs of camp on a sliding scale from $325 to $750 all-inclusive, depending on their economic circumstances. Some financial assistance is available, and YES! can supply you with an information packet designed to help you raise your own funds. Discounts are given to those who sign up for more than one camp or who get their friends to register. Camps are held from June through August; interested students are encouraged to apply well in advance.

You may want to call or write for the free *YES! Earth Action Guide,* an enthusiastic booklet that talks about what you can do in the areas of air and water pollution, rain forests, recycling, and so forth. The final page of the booklet has a long list of things your own YES! group could do to make a difference. Some of these are simple, like throwing a letter-writing party, while others involve more planning and motivation. If you want to ban the use of Styrofoam in your school or publish an article about student environmental efforts in the local paper, YES! can help make it happen. YES! welcomes student members with dues of $15 per year.

Youth for Environmental Sanity! also runs a two-week Sustainable Living Retreat in the summer, suitable for those ages sixteen to twenty-five who have already completed a YES! camp or who are otherwise already environmentally active. For more detailed information about specific programs and for application materials, contact YES!

Youth for Environmental Sanity! (YES!)
706 Frederick Street
Santa Cruz, CA 95062
Tel: 408-459-9344
Fax: 408-458-0255
Email: yes@cruzio.com
Web: http://www.yesworld.org

YOUTH IN NATURAL RESOURCES

College Course/Summer Study, Employment Opportunity

The Colorado Department of Natural Resources runs the Youth in Natural Resources program every summer. The program is designed to get high school students from groups that have not traditionally been represented in natural resource fields—the economically disadvantaged, ethnic minorities, females—to consider pursuing such careers. Each summer, Youth in Natural Resources places roughly one hundred and fifty high school students in full-time, paid positions with the Department of Natural Resources in locations all across Colorado. Participants not only work on such tasks as building trails and conducting wildlife inventories, but also complete a series of environmental lessons, keep a journal of their experiences, and go on field trips to explore other opportunities for careers in natural resources. They have the chance to spend three to five days on the campus of a Colorado college, during which they learn more about natural resources, tour the campus, and participate in a college planning workshop. Upon successfully completing the program, students receive a $100 savings bond to invest in their future educational pursuits.

The Colorado Department of Natural Resources also has about twenty-five openings for college students who wish to serve as crew leaders for Youth in Natural Resources participants. Contact the Colorado Department of Natural Resources for further details and information on application procedures.

Youth in Natural Resources
Colorado Department of Natural Resources
1313 Sherman Street, Room 718
Denver, CO 80203
Tel: 303-866-2540
Fax: 303-866-2115

Do It Yourself

FIRST

Imagine living in a city full of polluted air. The brown smog is so thick that it's difficult to breathe and see. The landfills are overflowing, and they're leaking chemicals into the groundwater that you drink. Sewage, chemicals, and solid waste are dumped into the ocean, destroying sea life and causing beaches to be closed. Animals and humans are becoming sick from chemical pesticides. Does this sound like a fictitious horror story? It's not. Each of these types of environmental problems has actually occurred somewhere in the United States.

We all know that many things need to be done to clean up the environment and prevent or cut back on further pollution. If you are considering a career in an environmental field, getting involved in environmental projects as a student is an excellent way to get valuable experience and give yourself a big head start in your chosen profession. How? By organizing and participating in projects such as cleanup days, water pollution monitoring, recycling, and so on, you have a chance to develop your skills in problem solving, organizing, teamwork, communications, leadership, and research. Another big plus is that you will build up experience—something all newcomers to the job market complain about: "They say I need experience to get the job, but how can I get experience if they won't give me a job?" Here's your chance. Create your own opportunities to get experience.

One thing most environmental workers share is a passion for the environment and a desire to get things done. This field is more like a calling than a job for some, and it takes dedication and the willingness to stand up for what you believe in, no matter what the opposition says. If you're the kind of person who sees a problem and says, "Somebody should do something about this," step aside. The environmental field needs people who see a problem and say, "What can *we* do about this?" Consider the following story.

Twelve-year-old Kory Johnson from Phoenix, Arizona, started a group called Children for a Safe Environment after her pregnant mother lost her baby when she drank polluted well water. Her group's first goal was to prevent a nearby incinerator from burning hazardous waste near her town. Kory wrote letters to the governor, local newspaper, department of environmental quality, and health department. To inform local residents, she mailed out countless letters and fliers. In addition, she spoke to the press about the issue at the incinerator site. Her group also held a protest at the state capital, as well as a candlelight rally, where members of her group spoke for three hours about protecting the environment.

As a result of Kory's efforts and the help of her mother, Greenpeace, and many volunteers, the governor announced that the incinerator project was off, even though the Arizona state government had already made a deal with the company to build it. Without Kory's strong commitment to getting the public involved in her cause, that incinerator might be burning hazardous waste today.

This is just one of many examples of young people who have taken the initiative and made a difference. Whatever Kory ends up doing in life, there's no doubt she'll be drawing on the experience she gained as a leader and organizer. Too often "young people feel very isolated and powerless," says Isabel Abrams, treasurer of Caretakers of the Environment, U.S.A. Don't. Too often, because they're still students living at home with their parents, young people believe their lives don't really begin until they're adults living on their own. Don't. If you see an environmental problem in your community, don't sit around and wait for somebody to do something about it; ask yourself, "What can *I* do about this?"

WHAT CAN YOU DO?

SCHOOL CLUBS

Many schools already have environmental clubs or groups for you to join. If your school does not, start one. You'll need a teacher to sponsor your club and, of course, you'll need at least two or three other students who are interested in working on environmental projects. Ask around school to find out who's interested. If there doesn't seem to be much interest, create some. Here's your first environmental problem: educating the public (your peers) about environmental issues. Most people are too busy doing what they're already interested in to pay attention to things they don't know much about. Your job is to get them involved by getting them concerned about something. How? Tell them about it between classes, at lunch, during homeroom. Write about it in the school newspaper, in your local paper, in your term paper. Take pictures and display these on posters or bulletin boards. It's much easier to generate interest for a club if you have an immediate problem to tackle. People are motivated right away and your chances of success are much better. Consider the following two examples:

> Miranda wants to start an environmental club because she likes nature and recycling and thinks it's really bad what they're doing to the rain forest. She talks to some friends at school, and the substitute math teacher agrees to be the sponsor. At their first meeting five people show up. Miranda calls the meeting to order and says, "So. Here's our environment club. What do we want to do?" Nobody really says much. They all just kind of look at each other until someone makes a joke and everybody bursts out laughing. There are a few attempts at bringing the discussion around to something environmental, but eventually everyone ends up talking about the upcoming football game and then it's time for the meeting to end.
>
> Veronica is disgusted with all the trash that litters the highway near her school. She talks to some friends about it and they agree it's an eyesore. So does the substitute math teacher. They decide to start an environmental club with the immediate goal of cleaning up the trash by the highway. They take some pictures and arrange to set up a small table in the lunch room during lunch hour where they display the pictures and talk about the project and the club to the people who wander by. At the first meeting twenty people show up. Veronica calls the meeting to order and says, "What are we going to do about this trash problem along the highway?" Everyone just kind of looks at each other until someone makes a joke and everybody bursts out laughing. Veronica laughs, too, then says, "No, seriously, what can we do?" One of the guys says, "We could organize a cleanup day," and then pretty soon everyone is talking. The meeting ends with the group setting a date for the cleanup day, appointing a committee to paint posters to advertise it, assigning a student to write an article about it for the school paper, and assigning another student to contact the state Adopt-a-Highway program.

Which club do you think is going to be more successful?

Write Letters

Robert F. Kennedy, Jr., environmental lawyer for the Natural Resources Defense Council, was once asked, "What's the most effective thing an individual can do to protect the environment?" Mr. Kennedy answered, "Write letters. Write personal letters to decision-makers. They keep a special file for personal letters, and they calculate that for every person who writes a personal letter, there are five thousand who hold the same opinion. So write letters!" Keep that in mind if you feel that your letters won't matter, because they will.

Contacting your member of Congress about environmental issues can be very effective, but before you do so there are some facts you need to know. There are two senators in your state and one representative in your congressional district, and you should choose your correspondent from these three people. This is because members of Congress traditionally think in terms of their responsibility to people who live in their geographic districts. Letters written to Congress members in districts other than your own will probably be sent back to your local representatives anyway. It is a good idea to contact the senator who belongs to the same political party as the president, because his or her office can often get things done faster and more efficiently. If both senators belong to the same party, you should contact the one with the most seniority.

To find out who your local representatives are, look in your local phone book or take a trip to the library. The library has many reference books about Congress, and the following reference book is an excellent source of more in-depth information on senators, local representatives, and congressional committees and subcommittees: *Congressional Staff Directory,* edited by P. Wayne Walker, available from CQ Staff Directories, Inc., Alexandria, Virginia. This directory is updated continually, so you should find the most recent edition. This book will give addresses, phone numbers, and other information for every Congress member in the country. For those of you with access to the Internet, similar Congressional information is available at many Web sites, among them The Great American Web Site, http://www.uncle-sam.com/, which bills itself as the Citizen's Guide to U.S. Government Resources on the World Wide Web

Once you know to whom to write, you need to determine *what* to write. A very important fact to note is that members of Congress take into consideration how much effort people put into contacting them. Therefore, you should write a personal letter because it means more than a form letter, phone call, telegram, or signatures on a petition. However, this does not mean that long letters are better. Letters that you write should be short (ideally only one page) and to the point. Your letter should be specific and informative, not rhetorical or bombastic. Congress member Morris Udall said once, "I usually know what the major lobbying groups are saying, but I don't often know of *your* experiences and observations, or what the proposed bill will do to or for *you*." Therefore, let your members of Congress know what the problem is, how it affects you, and what he or she can do to help. The most useful letters are factual accounts of your own personal experiences.

You should not send letters at the last minute, since it takes time for all of the decision-making to get done. Make sure you include your name and address, typed or printed, so your Representative or Senator can get back to you. If you do not hear from him or her in three to four weeks, send a follow-up letter. Finally, always remember that if your Congress member helped you or tried hard to help you, send him or her a note of appreciation.

Are you good at this? Maybe you're on your way to becoming an *environmental activist* or a *lobbyist*.

Plant Trees

Planting trees is a task with which many environmental groups are involved. For example, the Sierra Club often needs volunteers to help plant trees. On your own, you can plant trees around your school and, depending on the amount of land available, do much more with the landscape. For example, students at Jefferson High School in Rockford, Illinois, planted trees and shrubs along a drainage ditch and established a pine plantation and a prairie restoration area on school land.

Working on school property requires talking to officials in the school district to get permission. "The school district pretty much gave us free rein to plant and develop" on school land, said Bob Beebe, a science teacher from Jefferson High School. Next, you will need to get the trees, shrubs, seeds, and other materials that your planting project requires. Often, organizations may donate trees or sell them to you very cheaply. For example, the group at Jefferson High School received trees from a conservation service at no cost. If you can't get anyone to donate trees, you'll have to raise money to buy them. Is someone in your club artistic? Have him or her create a special design and put it on a T-shirt. Yard sales, bake sales, car washes, pet washes, carnivals, theater performances, music performances, and raffles are other common ways of raising money.

Are you good at this? Maybe you're on your way to becoming an *arborculturist, landscaper, grounds manager,* or *forest ranger.*

Buy Rain Forest Acres

The environmental club from Erindale Secondary School in Mississauga, Ontario, Canada, raised money to buy acres of rain forest land for $25 an acre. Each acre was then protected by the "Guardians of the Amazon" program, through the World Wildlife Fund. Their goal was to buy 75 acres, and they ended up raising enough money to adopt 280 acres. The club educated students in homeroom for a week, and then each homeroom tried to buy an acre, creatively raising the money to do so. Some of the fund-raising activities that took place included sending $2 singing telegrams to students in class, a teacher-student basketball game (that earned almost $1,000 in ticket sales), and a booth to sell hot dogs and hamburgers made with meat that did not come from former rain forest land. Students also bought their families Christmas presents of rain forest acres.

Are you good at activities like this? Maybe you're on your way to becoming a *fund-raiser.*

Clean Up

Cleaning up the environment is another project that students can undertake. You can approach this in a few different ways. First, there may already be a creek or river cleanup project taking place in your community that you can join. A good way to determine what projects are currently underway is to contact your state EPA, state or county department of public health, county soil and water conservation service, or department of environment. They will tell you what cleanup projects they themselves are working on, and may know what projects other groups are working on.

An example of a project started by a group outside city or state departments is a rivers project that originated at Southern Illinois University, Edwardsville. Honenegah High School in Rockton, Illinois, took part in this. The students collected data from one section of the river to determine how polluted it was. The college put this data into a database on the Internet, along with data from the other participants. Since each group of participants sampled different sections of the river, they created a valuable database regarding pollution levels in various areas of the river.

You can also start your own cleanup projects in your city. You will need to get permission from the city to do so, which should not be a problem. City officials are generally very cooperative with volunteers who are willing to clean up a polluted area of the city. Make sure to let them know if you have plans to plant any trees or shrubs on city land.

Another way to get involved in a cleanup is to participate in the Adopt-a-Highway program, which exists in many states. Through this program, your group can adopt a section of a state road to keep clean. You will need to call your state department of transportation for more information, since each state and each district within a state may have different procedures. Generally, the supervisor of your group will fill out an application to adopt a highway and request a specific state route that your group is interested in cleaning up. The section of adopted road is approximately two miles long, although it may be longer, and a sign is put up along the road with the name of your group. Your responsibility is to clean up the trash littered along your section of highway, which must be done once every one to two months from April to November. You must notify the Bureau of Maintenance at the Department of Transportation, so that they know when and where to pick up the bags of trash, which you leave along the road when finished.

Are you good at this? Maybe you're on your way to becoming a *ground-water professional, soil conservationist or technician, soil scientist,* or *pollution control technician.*

Recycle

Your school doesn't have a recycling program? Start one. You will need to seek permission from your principal. If your principal feels that you need to discuss the idea with your school board, you may have to present it to them. Generally, school officials have no problem with recycling programs, as long as they know your group will take care of the recyclables. You must present a clear, well-thought-out plan to both your principal and the board. Remember, you're handing them a solution to a problem, not another problem. Do some research on the Internet on what other schools have done to find out what works and what doesn't. For any recycling program to work, it must be just as easy for people to recycle as it is for them to do what they used to do with their trash. And they have to know about it. You'll need plenty of posters, school newspaper ads and articles, and announcements by teachers, as well as the example set by those who do recycle, to motivate the rest.

If your school already has a recycling program, see how much is really recycled. It may be that only soda cans and paper are recycled. The cafeteria is a large source of waste in schools, however. Plastic jugs, tin cans, glass containers, and milk cartons from the cafeteria can be recycled in many communities. If the cafeteria does not recycle, talk to the principal about changing this. Nine-year-old Brady Landon Mann, from Vancouver, Washington, for example, wrote a letter to his school principal requesting that their school start recycling when he found out that they did not. In return, he received a thank you note from the principal, and the school started to recycle. Simple as that. If a nine-year-old can do it, so can you.

While you're looking into cafeteria recycling practices, take a look at what type of trays and cups are used. Are they using Styrofoam or paper? If your school still uses Styrofoam, you can tackle the challenge of getting them to switch to paper. Often, Styrofoam is used because it's cheaper. Paper can be recycled, however, and it is biodegradable, making it worth a bit more expense. You will need to go to the principal again, and probably the school board, to request that paper be used instead of Styrofoam. Most likely you will run into some opposition here, because school boards are always concerned with cost. The board may not want to pay more for paper. However, if you can persuade them that the majority of students would rather pay a little extra for paper, you will have a much better chance of succeeding. To gain student support, educate your peers on the detrimental effects of Styrofoam on the environment, the recycling qualities of paper, and the cost difference between the two.

You can also take a look around the rest of your school to see which conservation practices have been implemented, and which ones need to be. So much paper is used in schools—by teachers, students, and office personnel—that it makes sense to use recycled paper. You can find out in the office if recycled paper is ordered by the school for teachers to use. If not, you can take on the project of getting the school to switch to recycled paper. You can also come up with ways to reduce the amount of paper used, and propose these ideas to the principal. In addition, you can examine the use of art supplies. Petroleum, a nonrenewable resource, is used in some paints and crayons. Some supplies also contain chemicals that are bad for your health when inhaled, such as toluene and ethanol. If your school uses supplies made with these chemicals, encourage them to use alternative supplies.

Are you good at this? Maybe you're on your way to becoming a *recycling coordinator.*

Save Energy

Something else you might want to find out is how much water and energy your school uses each year. If your school does not use low-flow toilets or water-reducing devices on faucets, for example, see if this can be changed. In addition, your school may not use fluorescent light bulbs, which are more energy-efficient than regular bulbs. Energy use can also be decreased by turning off lights and heating or air-conditioning when no one is in the building.

What about home energy use? Start a campaign to educate students about ways to save energy at home. Learn about insulation, thermostat settings, water use, light bulbs, efficient home heating and cooling systems, and windows and doors, and share your knowledge.

Are you good at this? Maybe you're on your way to becoming an *energy conservation technician.*

ON YOUR OWN

You don't necessarily have to be involved in a school club to get experience for a future environmental career. There is a lot that you can do as an individual to start an environmental program. One student, sixteen-year-old Dan Shuman from Dover, Pennsylvania, set up a creek testing program of his own. He came up with the idea after noticing that there were fewer numbers of certain species of fish in local lakes and streams. He was worried that acid rain may have caused the reduction in fish populations. He then learned from a fishing magazine that Dickinson College had a program that would provide kits to test water for acid rain for $20 each, and he decided to make it a Boy Scout project.

Dan started by asking local scoutmasters if boys could volunteer for the project. In return the Scouts would receive service hours. He then asked for Scout volunteers, and about fifteen wanted to participate. To raise money to buy the testing kits, he went to sports clubs, told them of the problem, and asked for donations. He received over $400 from three sports clubs for the project.

The group ran tests for a year on twenty-two streams. The data was sent to Dickinson College, where it was analyzed. The analysis confirmed Dan's theory that acid rain was contaminating the water in local streams. Dan informed the Scouts and sports clubs of the results, which prompted them to want to do more. Some of them wrote letters and called legislators. Although the legislators have not yet taken action, they all say that they will. All on his own, Dan identified a possible environmental problem. He did some research, persuaded others to join him, and together they identified a real environmental problem. Finally, they got the attention of those with the power to enforce a solution. Now even if Dan does not eventually become a *water pollution control technician,* he has gained valuable experience as a researcher, organizer, fundraiser, and leader—while he was doing something that genuinely interested him.

Sometimes it takes a little creative thinking to come up with a project idea. The effort paid off for Darlene Rodriguez, a senior at Miami Springs High

School. Latinos make up 60 percent of the students in Darlene's county, and many of them did not know much about environmental issues. It was important to her to get the message out to them and inspire some to get involved. Darlene put a brochure together in Spanish. Friends of the Everglades agreed to pay for the cost of the brochure, and some students took informal neighborhood surveys to determine what topics people would be interested in. Using the survey results, along with information about local environmental issues, Darlene wrote the brochure and distributed it. As a result, she provided information about the environment to a segment of the public that did not have it before. Maybe she won't go on to become a *journalist,* but she has certainly gained skills as a researcher and a reporter.

The yearbook staff at Trinity High School in Longview, Texas, wanted all students to be more involved in environmental issues. As a result, they made environmentalism the theme of their yearbook. They wrote articles about what students had done over the year to save resources, recycle, etc., and they printed the books on recycled paper. As a result, more students were inspired to become active in helping the environment, and the school eventually added an environmental science course.

CONCLUSION

We hope the examples in this chapter have persuaded you that, as a student, you can do something to help the environment right now. You can start programs where none existed before. You can educate your community about important environmental issues. You can motivate people to work together to solve environmental problems. In doing all this, you are combining something you are interested in—the environment—with actions that will build the skills and experience you will need in your future career.

III Surf the Web

First

You must use the Internet to do research, to find out, to explore. The Internet is the closest you'll get to what's happening now all over the place. This chapter gets you started with an annotated list of Web sites related to the environment. Try a few. Follow the links. Maybe even venture as far as asking questions in a chat room. The more you read about and interact with environmental personnel, the better prepared you'll be when you're old enough to participate as a professional.

One caveat: you probably already know that URLs change all the time. If a Web address listed below is out of date, try searching on the site's name or other key words. Chances are, if it's still out there, you'll find it. If it's not, maybe you'll find something better!

The Sites

The Center for Agriculture, Science, and Environmental Education

http://152.157.16.3/WWWSchools/OtherSites/CASEE/casee.html

CASEE is an innovative program in Washington's Battle Ground school district, which encourages high school students to create and work on month-long environmental projects for credit. What makes this program stand out is its emphasis on hard science and its close affiliation with environmental professionals as varied as master beekeepers, wildlife biologists, and highway transportation experts. One project described at length on-line was the CASEE Salmon Project in which students worked to improve creek conditions for Coho salmon. Each time a student completes research, he or she writes a "Big Picture" analysis, summing up how the project relates to the real world. Many of these can actually be read on-line. You can also read CASEE's monthly newsletters to get a glimpse into the breadth of projects. If you have a fledgling environmental program in your own school, this Web site could provide inspiration and reference resources.

CoolWorks

http://www.coolworks.com/showme

Can you picture yourself saddling up burros at the Grand Canyon or working as a tour guide at Mount Rushmore this summer? CoolWorks quickly links you up to a mass of information about seasonal employment at dozens of national and state parks, preserves, monuments, and wilderness areas. Most of the national and state jobs require that applicants be eighteen years or older. Specific job descriptions can also be accessed by clicking on a location that interests you on the USA Job Map. The map links to comprehensive descriptions of the park's attractions, facilities, and jobs currently available. You can also check out where recruiters for some of the larger parks and resorts will be visiting. If planning ahead isn't your forte, look to the "Help Wanted Now" listings for immediate job openings. While only a handful of jobs allow you to apply directly online, many have downloadable application forms.

Cyber-Sierra's Natural Resources Job Search

http://www.cyber-sierra.com/nrjobs/index.html

This site is overflowing with up-to-date information to assist job seekers in the fields of forestry, water resources, and other environmental careers. In spite of its abundance of data, the site has a friendly tone and seems to have the interests of a job seeker in mind. Take the time to read the Webmaster's Welcome page for insights into how to best use this site and general advice about looking for work. Even if you're only at the preliminary stage of exploring options for your future, there's plenty to glean from this site. Besides seeing the scope of career possibilities, you can link to many environmental organizations, online courses, and reference tools.

For students deciding on colleges or just what to do this summer, two areas could be of immediate value. The section called Forestry and Natural Resources Schools includes detailed information about many of the universities and colleges where you can earn a bachelor's degree in the environmental sciences. Another section called Summer Jobs is exactly what it sounds like: a list of numerous seasonal jobs and links to other job sites.

EE-LINK: ENVIRONMENTAL EDUCATION ON THE INTERNET

http://www.nceet.snre.umich.edu

Sponsored by the National Consortium for Environmental Education and Training, EE-Link aims to provide on-line educational resources for students and teachers interested in the environment. At this site, you'll find hundreds of links to environmental organizations, foundations, schools, and current projects that are on the Internet. A random sampling of links included the Costa Rica interactive multimedia class, the Coastal Marsh Project, and the Monarch Watch. Reference material is handily organized by topic, and there's an invaluable link to available grants from organizations you might not have found on your own.

There's definitely a bias toward educators here. In particular, the listings of conferences, workshops, jobs, and volunteer opportunities are mostly appropriate for teachers and college students. However, an enterprising high school student could probably carve out some possibilities with all that's available here.

ENN ONLINE: THE ENVIRONMENTAL SCIENCE SUPERSITE

http://www.enn.com

The Environmental News Network is a polished Web site with eye-catching photographs and graphics. (Because of all the images, a speedy modem will make visiting this site a lot more enjoyable.) This site—which functions as an on-line magazine, radio station, and discussion forum—provides timely updates and feature-length articles on environmental issues. In the "Newsroom" section, the latest environmental stories are covered in brief by seasoned scientists and journalists. "Planet ENN" provides lengthier articles, or you can tune in to "ENN Radio" by downloading the Real Audio plug-in. Budding journalists could even make a contribution to ENN by submitting an article to its editor; the guidelines are posted on-line. A list of other environmental sites on the Web looked promising, but many of its links were outdated. Career information per se is limited, but if you're trying to stay conversant on a wide range of environmental concerns, you'll want to bookmark ENN for regular reads.

ENVIRONMENTAL CAREERS ORGANIZATION

http://www.eco.org

Founded in 1972 as an experimental internship program in Massachusetts, the Environmental Careers Organization (ECO) has developed into a national, nonprofit organization offering career advice, career products, and numerous environmental internships across the country. If you're looking for an environmental internship—for example, assisting with noxious weeds or timber surveys—this is the place to go. The site provides a quick glimpse at the organization's history, its programs (including an annual conference on career opportunities) and a detailed list of available internships. Lasting from a few weeks in the summer to more than a year, the internships are mostly for college students or college graduates. You can send your resume and apply for internships electronically at the site.

This site is well suited to people with a college education who know what they're looking for. But descriptions of internships should help high school students plan ahead, and the site provides links with other career pages and environmental organizations.

ENVIRONMENTAL PROFESSIONAL'S HOMEPAGE

http://www.clay.net

Just the facts—that's what ClayNet's site has to offer. This home page for environmental professionals provides a straightforward index to a wealth of factual data and information. Sponsored by GZA GeoEnvironmental Technologies, ClayNet isn't interested in dazzling graphics or in putting a spin on environmental news. Instead, its mission statement to provide "hard-core data retrieval" actually holds up. It's fast, reliable, and will link you to original source documents from the EPA and other agencies.

If you're searching for a professional association in your area, Clay Net lists them in categories that include Environmental/Ecological, Life Sciences, Earth Sciences, Chemistry, and others. If you need reference information, recent software tools, or the address of a government agency, ClayNet is an excellent place to look. This site also has direct links to the latest regulations, legislation, and professional conferences. While you won't find information specifically geared toward students, the bulletins and announcements area could provide a way to approach environmental professionals about internship possibilities.

GIVE WATER A HAND

http://www.uwex.edu/erc/

This program, sponsored by the University of Wisconsin's Environmental Resources Center, encourages young people to take leadership roles in water protection efforts in their communities. The site's centerpiece is an extensive *Action Guide* that can be downloaded chapter by chapter, according to individual needs. These chapters include information on how to map a watershed, how to ask an expert for help, and how to keep the project on track. If you're just starting out, click on your location in the U.S. map to find the helpful Give Water a Hand coordinator in your state. You can also link to related educational materials and to other organizations involved in watershed efforts.

GLOBAL LEARNING AND OBSERVATIONS TO BENEFIT THE ENVIRONMENT (GLOBE)

http://www.globe.gov/ghome/invite.html

GLOBE is an ambitious project in which students from around the world take core sets of scientific measurements and observations about their local environment, such as air temperature, precipitation, and soil moisture. The data is forwarded to GLOBE-affiliated scientists who record it on-line for all participants to access. While there's plenty of fun activities like an on-line expedition to Mt. McKinley, GLOBE emphasizes learning solid scientific protocol. Visit the Scientist's Corner to read scientific research papers or to see how data is used to create global contour maps. Students on every continent are taking part in GLOBE, and students from a high school in Arkansas might find themselves exchanging data with students in Finland or China. One complaint is that the site seems to have grown quickly, without enough thought given to organizing the site for more expedient visits.

GLOBAL RIVERS ENVIRONMENTAL EDUCATION NETWORK (GREEN)

http://www.igc.apc.org/green/

If the quality of water in your community is a source of concern, this site might move you toward solving the problem. GREEN is dedicated to developing and disseminating resources to students and teachers who want to study and improve the "health" of their local watersheds. The educational model is action-oriented and simple: students monitor the quality of the water, evaluate specific changes or larger trends, and if necessary, take action to improve the watershed's sustainability. Students at North Farmington High School near Detroit, for example, detected increased levels of bacterial contaminants downriver from a pipe that exited a city sewage pumping station. They reported their findings to the city engineer, who acted quickly to repair a malfunctioning pump. At this site, you'll find on-line conferences where GREEN participants share problems, tips, and ideas about water monitoring. There's even a "student lounge" where students can exchange general information about themselves.

THE INTERACTIVE FROG DISSECTION

http://teach.virginia.edu/go/frog/

Dissect a frog on-line? If you've got the amphibian, a scalpel, and an Internet connection, you're ready. With color photographs and downloadable QuickTime movies, this Web site will call to mind the aroma of formaldehyde. The site was designed to provide a step-by-step interactive tutorial for use in high school biology classrooms. It can also be used as a preparation tool or as a substitute for an actual laboratory dissection. The complete dissection is divided into Introduction, Preparation, Skin Incisions, Muscle Incisions, and Internal Organs. In each section, procedures for dissection are clearly described and presented in photographs. Anyone interested in biology will want to check out this site, as will students who are thinking of authoring their own tutorial Web site.

INTERNATIONAL EDUCATION AND RESOURCE NETWORK (i*EARN)

http://www.igc.apc.org/iearn/

i*EARN provides a wonderful site for students who feel the tug of international humanitarian and environmental causes. This organization strives to help teachers and students work with people in other parts of the world through a low-cost telecommunications network. At the Web site, there's a Youth Action page that lists global projects needing participants and gives specific actions you can take. Recent projects have included Project Hope in China, UNICEF's Voices of Youth, and help for flood victims in Sabah, Malaysia. A semiannual newsletter called *Planetary Notions* gives students an opportunity to publish their own articles in English, Russian, and Spanish. There's also a purely educational aspect to i*EARN. During the Sahara to Antarctica project, for example, two experienced explorers are posting their journal entries on-line in order to be the eyes and ears for students and teachers around the world. i*EARN's site would be improved if it were segmented into age groups; certain areas are too juvenile for a high school audience, while others are clearly aimed at teachers. But i*EARN, with business partners like IBM, the Discovery Channel, Com monwealth Edison, and the American Red Cross, offers a great way to connect on a global scale with peers who care about the future of the world.

INTERN-NET

http://www.vicon.net/~internnet

Intern-NET is a new database for students seeking internships in certain fields. From the home page, you can search for an internship by selecting a category (i.e., parks/outdoor recreation or environmental interpretation/education), then choosing a region of the United States where you want to work. A search for environmental interpretation/education jobs in the southern United States, for example, turned up internships with the Nature Conservancy in Tennessee, the Marine Resources Development Foundation in Florida, and Nature's Classroom on Lookout Mountain in Alabama, to name a few. Most of these provided stipends, housing, and meals. Students can also apply for a $200 scholarship for any job they locate through the database.

Outdoor Action Guide to Outdoor/Environmental Careers

http://www.princeton.edu/`oa/careeroe.html

If you're only beginning your search for an environmental career, this short but useful home page is a good starting place. The material was written as a resource guide for a Princeton University workshop on outdoor and environmental careers, but many people will be interested in its contents: finding your career, types of careers, career examples, job newsletters, books, other resources, graduate programs in environmental education, environmental job leads, basic job questionnaire, and trends in employment. The job questionnaire is especially helpful for defining the type of environmental job you want. Although no subject is discussed in great detail, the guide does function as a one-stop shopping area for information on environmental careers. If you want a list of twenty-five books on environmental jobs, or if you want a list of important environmental publications, you'll find it here. Because this home page is short, you might want to print out an old-fashioned hard copy.

The Princeton Review

http://www.review.com/

This site is everything you want in a high school guidance counselor—it's friendly, well informed, and available to you night and day. Originally just a standardized test preparation company, the Princeton Review is now an online source giving you frank advice on colleges, careers, and of course, SATs. Students who've spent their summers and after-school hours working for environmental causes will find good tips here on how to present those extracurricular activities on college application forms. There's a handy link to *Time*'s interactive guide, "The Best College for You," which discusses costs, admissions, and alternatives to four years of college. If you want to contact other students who are also weighing their options, link to one of the discussion groups on college admissions and careers. Two of the Princeton Review's coolest tools are the Find-O-Rama, which creates a list of schools based on the criteria you type in, and the Counselor-O-Matic, which reviews your grades, test scores, and extracurricular record to calculate your chances of admission at many colleges.

Rainforest Action Network (RAN)

http://www.ran.org/

The Rainforest Action Network has a Web site that you don't merely visit: you make a statement while you're there. This dynamic, colorful home page draws you in with its sense of urgency. For instance, click on Action Alert or Urgent to read about current issues affecting the rain forest. But don't just read about them—show your support by sending an email to legislators right from the site, using RAN's sample letters or customizing them. The Campaigns section describes ongoing campaigns such as the Amazon program, the Ancient Redwood Campaign, and the Protect an Acre program, and the Victories section keeps track of RAN's successful campaigns back to 1987. If spending time at this site only whets your appetite for more and you're willing to spend a summer in San Francisco, you can check out the numerous volunteer and internship possibilities that are listed on-line.

School Nature Area Project

http://www.stolaf.edu/other/snap/

Schools across Minnesota joined St. Olaf College to create the School Nature Area Project (SNAP). Each school selects a nearby nature area, determines its environmental needs, and creates a site management plan. As a result, students are active participants in planting prairies, creating butterfly sanctuaries, restoring wetlands, and developing trails in their home communities. The online *SNAPSHOTS* newsletter highlights these achievements. The newsletter is organized by seasonal projects going back to 1994. Dig into each section, and you'll find valuable ideas on how these programs are integrated into the curriculum. Another area of the SNAP site details how students and teachers can apply for individual grants for more personal projects that enhance native vegetation and wildlife habitat. This site is worth visiting not only to get ideas for your own projects, but to see the vast impact that students can have on their communities.

Sierra Student Coalition

http://www.ssc.org/

The Sierra Student Coalition (SSC) is the offspring of the mother of all environmental organizations, the Sierra Club. While much of SSC's site was under construction at the time of our visit, it holds strong promise for interactivity. For instance, members are encouraged to go beyond being passive Web site browsers and become publishers on the site. "Project Local Space" provides templates where students can insert their regional group's information. In addition to just reading about current Sierra Club campaigns on-line, you can instantly email your feedback to Congress, the White House, or government officials from the site. Other sections help you find local group and contact information, request resources for running regional campaigns, and tap into the "Idea Bank" to learn from the experiences of other student activists. In addition, *generation E,* the SSC newsletter, is coming online soon.

STUDENT CONSERVATION ASSOCIATION

http://www.sca-inc.org/

If the Student Conservation Association (SCA) isn't a familiar name to you already, it should be. Dedicated to "changing lives through service to nature," this group's Web site is bursting with over one thousand listings of volunteer and internship positions across the country. This site makes it almost too easy: you can access a list of current positions, get information, and request an application all on-line. Via the Web, you can also get subscription information for *EarthWork*, a publication on conservation issues with a popular, comprehensive "JobScan" listing. Several other resource guides and books can also be ordered on-line.

UBIQUITY

http://ourworld.compuserve.com/homepages/ubikk

Here's a friendly site that was created with the sole purpose of introducing students to the environmental job market. The most frequently visited page is the Environmental Job Descriptions Page, which provides brief job summaries for over twenty (mostly entry-level) jobs, including environmental regulatory specialist, occupational health nurse, hazardous materials technician, and environmental engineer. Elsewhere in Ubiquity's site, you'll find resume and cover letter writing tips geared specifically at job seekers in environmental fields. You can search for an internship from this site, too.

Read a Book

First

When it comes to finding out about the environment, don't overlook a book. (You're reading one now, after all.) What follows is a short, annotated list of books and periodicals related to the environment. The books range from fiction to personal accounts to biographies of the greats. Don't be afraid to check out the professional journals, either. The technical stuff may be way above your head right now, but if you take the time to become familiar with one or two, you're bound to pick up some of what is important to environmental personnel, not to mention begin to feel like a part of their world, which is what you're interested in, right?

We've tried to include recent materials as well as old favorites. Always check for the most recent editions, and, if you find an author you like, ask your librarian to help you find more. Keep reading good books!

Books

Ausubel, Ken. *Restoring the Earth: Visionary Solutions from the Bioneers.* Tiburon, CA: H. J. Kramer, 1997. The founder of the Bioneers Conference, an annual event for ecologists from around the world, presents profiles of visionary men and women and their solutions for the planet's most pressing ecological problems.

Barr, Nevada. *Endangered Species.* New York: Putnam, 1997. Latest in a series of ecological mysteries featuring Anna Pigeon, park ranger. In the Cumberland Island National Seashore in Georgia, in a world of loggerhead turtles and wild ponies, Anna finds herself embroiled in a plot involving espionage, airplane crashes, eccentric islanders, and many twists and turns of plot before she can unravel the mystery.

Brisk, Marion A. *1001 Ideas for Science Projects on the Environment.* New York: Arco, 1997. These practical and original project ideas cover all the environmental issues of greatest concern to today's young people and are sure to spark the imagination of students from middle school to college level.

Dashefsky, H. Steven. *Kids Can Make a Difference!:* Environmental Science Activities. Blue Ridge Summit, PA: TAB Books, 1995. An overview of pollution on land and sea and what you can do now to create positive change.

Dee, Catherine, editor. *Kid Heroes of the Environment: Simple Things Kids Are Doing to Save the Earth.* Berkeley, CA: Earthworks Press, Inc., 1991. Profiles of twenty-five young people, ages seven to seventeen, who have made positive contributions to environmental concerns both locally and nationally.

Dickinson, Peter. *Eva.* New York: Dell, 1990. After a terrible accident, Eva wakes to discover that her brain has been transplanted into the body of a chimpanzee. She becomes a champion for chimps in captivity after learning what it means to think and feel like an animal in the novel set slightly in the future.

Dwyer, Jim. *Earth Works: Recommended Fiction and Non-fiction About Nature and the Environment for Adults and Young Adults.* New York: Neal-Schuman, 1996. Excellent bibliographic source for books that will help individuals develop the knowledge and understanding that can lead to a commitment to environmental causes. Lists materials in each section for adults and for teens.

Education for the Earth: A Guide to Top Environmental Studies Programs. Princeton, NJ: Peterson's Guides, 1993. An overview of various environmental careers and profiles of the best undergraduate programs in environmental studies.

Ehrenfeld, David. *Beginning Again: People and Nature in the New Millennium.* Oxford, England: Oxford University Press, 1995. A collection of twenty quick, convincing essays on our relationship to each other and the planet recounting how greed, ignorance, and pride have put us in a precarious situation that must be reversed.

Ehrlich, Paul and Anne Ehrlich. *Betrayal of Science and Reason: How Anti-Environmental Rhetoric Threatens Our Future.* Washington, DC: Island Press, 1997. Hard-hitting and timely book in which two world-renowned scientists speak out against the efforts of anti-environmental forces and politicians to downplay the seriousness of conservation issues. An essential study for any student interested in ecology as a career.

Feldman, Andrew J. *Sierra Club Green Guide: Everybody's Desk Reference to Environmental Information.* San Francisco: Sierra Club, 1996. More than twelve hundred information sources, organizations, services, and Internet sites covering environmental issues with a section on "green" living, shopping, travel, and employment opportunities.

Galdikas, Birute. *Reflections of Eden: My Years with the Orangutans of Borneo.* Boston: Little, Brown & Co., 1995. A protégé of Louis Leakey, Galdikas went to Indonesia in 1971 to study the orangutan and has been there ever since. Her groundbreaking scientific and conservation work has been the subject of numerous articles and more than twelve TV documentaries.

Gartner, Robert. *Careers Inside the World of Environmental Science.* New York: Rosen, 1995. Excellent descriptions of a variety of jobs that will contribute to positive change for our environment.

Gartner, Robert. *Exploring Careers in the National Parks.* New York: Rosen, 1997. Offers career tips and training information for many jobs available in National Parks throughout the country.

Heyerdahl, Thor. *Green Was the Earth on the Seventh Day.* New York: Random House, 1996. Warm, spirited, and amusing memoir of Heyerdahl's youth, setting out with his new wife to discover the natural and unspoiled world of the South Pacific. This was the first of many journeys and expeditions that would lead to his distinguished career as a naturalist and adventurer.

Klass, David. *California Blue.* New York: Scholastic, 1994. When seventeen-year-old John Rodgers discovers a new subspecies of butterfly that may necessitate closing the lumber mill where his dying father works, he and his father find themselves on opposite sides of the environmental conflict. A gripping novel that brings to life the heated emotions on both sides of an environmental issue.

Krensky, Stephen. *Four Against the Odds: The Struggle to Save Our Environment.* Boston: Houghton, Mifflin, 1992. Exciting biographical accounts of four major activists who spearheaded environmental reform: John Muir and the creation of National Parks, Lois Gibb's fight for victims of toxic waste in New York's Love Canal, Rachel Carson's revelations of the dangers of DDT, and Chico Mendes' crusade to save South American rain forests.

Krupin, Paul. *Krupin's Toll-Free Environmental Directory.* Kennewick, WA: Direct Contact Publishing, 1994. Comprehensive nationwide listing of over 4,500 environmental firms, organizations, government agencies, and private institutions of interest to career seekers and working professionals in the field.

Lear, Linda J. *Rachel Carson: Witness for Nature.* New York: Henry Holt & Co., 1997. Definitive biography of a woman who combined a scientific background with a clear and passionate writing style to alert the masses to the dangers of pollution.

Lisowski, Marylin and Robert A. Williams. *Wetlands.* New York: Franklin Watts, 1997. Projects and activities that explore five major types of wetlands and demonstrate why they are essential to our life on earth.

McVey, Vicki. *The Sierra Club Kid's Guide to Planet Care & Repair.* San Francisco: Sierra Club, 1993. Includes stories of young people who have made a difference in preventing further damage to the delicate balance of nature and humanity.

Montgomery, Sy. *Walking with the Great Apes: Jane Goodall, Dian Fossey, Birute Galdikas.* Boston: Houghton Mifflin, 1991. Biography of three women who have devoted their lives and work to working with animals in the wild rather than in zoos and laboratories. They are all fine examples of ecofeminism in the field.

Muir, John. *A Thousand-Mile Walk to the Gulf: The travels of noted conservationist and naturalist John Muir.* San Francisco: Sierra Club, 1991. Muir's hike from Kentucky to Florida's Gulf Coast occurred soon after the Civil War and focused on nature and conservation as well as his social observations on a war-torn land and its people.

Pringle, Laurence. *Scorpion Man: Exploring the World of Scorpions.* New York: Scribner's, 1994. Profile and photo-essay of the work of wildlife biologist Gary Polis, who specializes in the study of scorpions, including forces in his life that led to his choice of career. See also: *Jackal Woman* (New York: Scribner's, 1993) and *Dolphin Man* (New York: Atheneum, 1995) by the same author for profiles of similar wildlife biologists.

Quintana, Debra. *100 Jobs in the Environment.* New York: Macmillan, 1997. Explores a wide variety of work from marine animal trainer to nature preserve manager to lobbyist and fund-raiser.

Robinson, Kim Stanley. *Blue Mars.* New York: Bantam, 1997. Final volume of a trilogy that includes *Red Mars* and *Green Mars.* Colonists who have transformed Mars into a perfectly habitable world find their work complicated by impending ecological disaster on Earth and political strife between the "Reds" who want to preserve Mars in its desert state and the "Greens" who plan to transform the planet into a version of Earth.

Schueler, Don. *A Handmade Wilderness.* Boston: Houghton Mifflin, 1996. Suspenseful, funny, and deeply moving account of two men, one white and one black, who in 1968 purchased eighty acres in Mississippi to bring back the native plant and animal life, creating a wilderness area that contained every ecosystem in the region.

Sharp, Bill (principal author). The Environmental Careers Organization. *The New Complete Guide to Environmental Careers.* 2nd edition. Washington, DC: Island Press, 1993. Details on many career opportunities, including background information, issues and trends, salary, and resources to contact for further information. Focus is on people with profiles and case studies of actual environmental professionals in action. (A 3rd edition is planned for November 1998. Call Island Press at 202-232-7933 for information.)

Shepard, Paul, editor. *The Only World We've Got: A Paul Shepard Reader.* San Francisco: Sierra Club, 1996. A collection of essays by a man considered to be an elder in the environmental movement, ranging in content from human ecology to our relationships with our fellow animals.

Stefoff, Rebecca.*The American Environmental Movement.* New York: Facts on File, 1995. Chronological approach that begins with the lifestyles of Native Americans and continues through the development of American industry and the rise of the conservation movement.

Taylor, Theodore. *The Bomb.* San Diego: Harcourt, Brace & Co., 1995. A sixteen-year-old native of Bikini Atoll in the South Pacific leads the protest against the U.S. government using his ancestral homeland as a site for atomic bomb testing in the period following World War II. Poignant novel based in part on the author's experiences in the U.S. Navy.

Thomas, Peggy. *Medicines from Nature.* New York: Twenty-first Century, 1997. Fascinating account of the work of ethnobotanists, who search the oceans and rain forests for potential medicines to cure deadly diseases. Includes discussion of the threats to biodiversity and its consequences.

Werbach, Adam. *Act Now, Apologize Later.* New York: HarperCollins, 1997. Werbach's activist career started at age eight when he organized a petition drive at his school. At age thirteen he founded the Sierra Student Coalition which now has thirty thousand members, and he went on to become the youngest president the Sierra Club has ever had. This is his call to action against the erosion of environmental laws.

Wilson, Edward O. *Naturalist.* Washington, DC: Island Press, 1994. Memoir of one of the most prominent biologists working today, in which he celebrates the changes in our view of nature in this century—namely, that we know now "we are bound to the rest of life in our ecology, our physiology, and even our spirit."

PERIODICALS

E: the environmental magazine. Earth Action Network, 28 Knight Street, Norwalk, CT 06851. Bimonthly. Addresses ecology and environmental protection within the framework of citizen participation and public policy.

Environmental Careers Bulletin. Environmental Careers Bulletin, 11693 San Vicente Boulevard, Suite 327, Los Angeles, CA 90049, 310-399-3533. Monthly. Opportunities from a wide selection of environmental occupations.

Environmental Law. American Bar Association, Standing Committee on Environmental Law, 1800 M Street, NW, Washington, DC 20036, 202-331-2276. Twice annually. Presents information about the field of environmental litigation. This publication is free to individuals.

Environmental Protection Magazine. Stevens Publishing Corporation, 225 North New Road, Waco, TX 76710, 817-776-9000. Monthly. Dedicated to air and water pollution control.

EPA Journal. U.S. Environmental Protection Agency, 401 M Street, SW, A-107, Washington, DC 20460, 202-382-4393. Bimonthly. Includes air, water, and noise pollution issues.

The Ground Water Age. National Trade Publications, Inc. 13 Century Hill, Latham, NY 12110-2197, 518-783-1281. Monthly. A publication geared toward groundwater professionals.

Hazardous Substances & Public Health. U.S. Department of Health & Human Services, Toxic Substances/Disease Regulations, 1600 Clifton Road NE, E-33, Atlanta, GA 30333, 404-639-0746. Bimonthly. Connects hazardous waste issues and human health.

Hazardous Waste Business. McGraw-Hill, 1221 Avenue of the Americas, New York, NY 10020, 212-512-2000. Biweekly. General news concerning hazardous waste.

Hazardous Waste Management Guide. J. J. Keller & Associates, Box 368, Neenah, WI 54957, 414-722-2848. Semiannual. Management guidelines for dealing with hazardous waste, from identifying to disposing.

Journal of Forestry. (Forestry Quarterly) Society of American Foresters, 5400 Grosvenor Lane, Bethesda, MD 20814, 301-897-8720. Monthly. Discusses numerous aspects of forest use, including protection and management.

National Parks: the Magazine of the National Parks & Conservation Association. National Parks & Conservation Association, 1776 Massachusetts Avenue, NW, Washington, DC 20036. Bimonthly. Aimed at the general public. Includes information about national parks and reserves of the United States and conservation of natural resources.

National Wildlife. National Wildlife Federation, 8925 Leesburg Pike, Vienna, VA 22184. Bimonthly. A popular magazine devoted to wildlife conservation issues.

Ocean Development and International Law: The Journal of Marine Affairs. Taylor & Francis, Inc., 1900 Frost Road, Suite 101, Bristol, PA 19007-1598, 215-785-5800. Bimonthly. Discusses current issues involving oceans worldwide.

Pollution Engineering: Magazine of Environmental Control. Pudvan Publishing Company, PO Box 5080, Des Plaines, IL 60061-5080, 708-635-8800. Thirteen times per year. A trade publication for pollution control professionals and environmental engineers.

Recycling Today. G. I. E., Inc., 4012 Bridge Avenue, Cleveland, OH 44113, 216-961-4130. Monthly. A newsletter that serves the recycling industry.

Sierra. Sierra Club, 730 Polk Street, San Francisco, CA 94109. Monthly (except bimonthly July-August and November-December). Major concerns are environmental protection and conservation of natural resources.

Waste Management: Nuclear, Chemical, Biological, Municipal. Pergamon Press, Inc., 660 White Plains Road, Tarrytown, NY 10591, 914-524-9200. Quarterly. International in scope, this magazine presents information covering all aspects of waste and its disposal.

III

Ask for Money

First

By the time most students get around to thinking about applying for scholarships, they have already extolled their personal and academic virtues to such lengths in essays and interviews for college applications that even their own grandmothers wouldn't recognize them. The thought of filling out yet another application form fills students with dread. And why bother? Won't the same five or six kids who have been fighting over grade point averages since the fifth grade walk away with all the really *good* scholarships?

The truth is, most of the scholarships available to high school and college students are being offered because an organization wants to promote interest in a particular field, encourage more students to become qualified to enter it, and finally, to help those students afford an education. Certainly, having a good grade point average is a valuable asset, and many organizations that grant scholarships request that only applicants with a minimum grade point average apply. More often than not, however, grade point averages aren't even mentioned; the focus is on the area of interest and what a student has done to distinguish himself or herself in that area. In fact, frequently the *only* requirement is that the scholarship applicant must be studying in a particular area.

Guidelines

When applying for scholarships there are a few simple guidelines that can help ease the process considerably.

Plan Ahead

The absolute worst thing you can do is wait until the last minute. For one thing, obtaining recommendations or other supporting data in time to meet an application deadline is incredibly difficult. For another, no one does their best thinking or writing under the gun. So get off to a good start by reviewing scholarship applications as early as possible—months, even a year, in advance. If the current scholarship information isn't available, ask for a copy of last year's version. Once you have the scholarship information or application in hand, give it a thorough read. Try and determine how your experience or situation best fits into the scholarship, or even if it fits at all. Don't waste your time applying for a scholarship in literature if you couldn't finish *Great Expectations*.

If possible, research the award or scholarship, including past recipients and, where applicable, the person in whose name the scholarship is offered. Often, scholarships are established to memorialize an individual who, for example, majored in religious studies or loved history, but in other cases, the scholarship is to memorialize the *work* of an individual. In those cases, try and get a feel for the spirit of the person's work. If you have any similar interests or experiences, don't hesitate to mention these.

Talk to others who received the scholarship, or to students currently studying in the same area or field of interest in which the scholarship is offered, and try to gain insight into possible applications or work related to that field. When you're working on the essay asking why you want this scholarship, you'll have real answers—"I would benefit from receiving this scholarship because studying engineering will help me to design inexpensive but attractive and structurally sound urban housing."

Take your time writing the essays. Make certain you are answering the question or questions on the application and not merely restating facts about yourself. Don't be afraid to get creative; try and imagine what you would think of if you had to sift through hundreds of applications. What would you want to know about the candidate? What would convince you that someone was deserving of the scholarship? Work through several drafts and have someone whose advice you respect—a parent, teacher, or guidance counselor—review the essay for grammar and content.

Finally, if you know in advance which scholarships you want to apply for, there might still be time to stack the deck in your favor by getting an internship, volunteering, or working part-time. Bottom line: the more you know about a scholarship and the sooner you learn it, the better.

Follow Directions

Think of it this way: many of the organizations who offer scholarships devote 99.9 percent of their time to something other than the scholarship for which you are applying. Don't make a nuisance of yourself by pestering them for information. Follow the directions given to you, even when asking for the application materials. If the scholarship materials specify that you write for further information, write for it—don't call.

Pay close attention to whether you're applying for an award, a scholarship, a prize, or some other kind of financial aid. Often these words are used interchangeably, but just as often they have different meanings. An award is usually given for something you have done: built a park or helped distribute meals to the elderly; or something you have created: a design; an essay; a short film; a screenplay; an invention. On the other hand, a scholarship is frequently a renewable sum of money that is given to a person to help defray the costs of college. Scholarships are given to candidates who meet the necessary criterion based on essays, eligibility, or grades, and sometimes all three.

Supply all the necessary documents, information, and fees, and make the deadlines. You won't win any scholarships by forgetting to include a recommendation from your history teacher or failing to postmark the application by the deadline. Bottom line: get it right the first time, on time.

Apply Early

Once you have the application in hand, don't dawdle. If you've requested it far enough in advance, there shouldn't be any reason for you not to turn it well in advance of the deadline. You never know, if it comes down to two candidates, the deciding factor just might be who was more on the ball. Bottom line: don't wait, don't hesitate.

Be Yourself

Don't make promises you can't keep. There are plenty of hefty scholarships available, but if they all require you to study something that you don't enjoy, you'll be miserable in college. And the side effects from switching majors after you've accepted a scholarship could be even worse. Bottom line: be yourself.

Don't Limit Yourself

There are many sources for scholarships, beginning with your guidance counselor and ending with the Internet. All of the search engines have education categories. Start there and search by keywords, such as "financial aid," "scholarship," "award." Don't be limited to the scholarships listed in these pages.

If you know of an organization related to or involved with the field of your choice, write a letter asking if they offer scholarships. If they don't offer scholarships, don't let that stop you. Write them another letter, or better yet, schedule a meeting with the president or someone in the public relations office and ask them if they would be willing to sponsor a scholarship for you. Of course, you'll need to prepare yourself well for such a meeting because you're selling a priceless commodity—yourself. Don't be shy, be confident. Tell them all about yourself, what you want to study and why, and let them know what you would be willing to do in exchange—volunteer at their favorite charity, write up reports on your progress in school, or work part-time on school breaks, full-time during the summer. Explain why you're a wise investment. Bottom line: the sky's the limit.

THE LIST

AGI Minority Participation Program Scholarship
American Geological Institute
4220 King Street
Alexandria, VA 22302
Tel: 703-379-2480

Open to African-Americans, Native Americans, and Latinos with a major in environmental studies. Fifty awards averaging $1,500 are given each year.

AMS Industry Undergraduate Scholarships
American Meteorological Society
45 Beacon Street
Boston, MA 02108-3693
Tel: 617-227-2426 ext 235

To encourage outstanding undergraduate students to pursue careers in the atmospheric and related oceanic and hydrologic sciences. Students in the following fields are encouraged to apply: atmospheric sciences, oceanography, hydrology, chemistry, computer sciences, engineering, environmental sciences, mathematics, and physics. Candidates must have a minimum GPA of 3.0, and must be U.S. citizens or permanent residents. The American Meteorological Society (AMS) encourages applications from women, minorities, and students with disabilities. Awards are based on merit and potential for accomplishment in the field. There are twelve awards worth $2,000 each given every year.

Argonne National Laboratory
Division of Education Program
9700 South Cass Avenue
Argonne, IL 60439
Tel: 708-252-2000

Student research participation program and thesis research; open to undergraduate and graduate students. Must be U.S. citizen. Participants receive a stipend of $225 per week and complimentary housing or a housing allowance. Transportation expenses are reimbursed for one round trip between Argonne and the participant's home or university for distances greater than one hundred miles.

Charles A. Lindbergh Fund
708 South 3rd Street, Suite 110
Minneapolis, MN 55415-1141
Tel: 612-338-1703

Citizens of all countries are eligible who are interested in the advance of technology and in the preservation of the environment. The $10,580 award is given to up to ten recipients annually.

Clara Carter Higgins Scholarship
The Garden Club of America
598 Madison Avenue
New York, NY 10022
Tel: 212-753-8287

Must be currently studying, or planning on studying: Environmental Studies, Ecology. Award is meant to be used to cover tuition for summer course(s) in Environment Science/Environmental Studies. The award is approximately $500; the number of scholarships available varies.

Culturally Diverse Institutions Undergraduate Student Fellowships
Environmental Protection Agency
401 M Street, SW
Washington, DC 20460
Tel: 800-490-9194

Applicants for this program must be U.S. citizens or permanent residents who are enrolled full-time with a minimum GPA of 3.0 in a four-year accredited institution that meets the definition of the Environmental Protection Agency (EPA) as a culturally diverse institution. Students must be majoring in environmental science, physical sciences, biological sciences, computer science, environmental health, social sciences, mathematics, or engineering. They must be available to work as interns at an EPA facility during the summer between their junior and senior years. The fellowship provides payment of tuition and fees, an annual book allowance of $250, and an annual stipend of $1,125. During the summer internship, students receive up to $600 in relocation support and up to $5,000 as a stipend to cover living expenses.

Earthwatch Scholarship and Fellowship Program
Earthwatch Expeditions
680 Mount Auburn Street, PO Box 403
Watertown, MA 02172
Tel: 617-926-8200

Open to high school juniors and seniors; national competition for high school students and teachers only. Two hundred awards worth an average of $2,000 each are given each year.

Environmental Sciences Division Scholarship
American Nuclear Society
555 North Kensington Avenue
La Grange Park, IL 60525
Tel: 708-352-6611

Eligible to apply are undergraduate students enrolled in a nuclear science or nuclear engineering program at an accredited institution in the United States. They must have completed at least two academic years, be U.S. citizens or permanent residents, and be able to demonstrate academic achievement. Applicants must be sponsored by an American Nuclear Society local section, division, student branch, committee, member, or organization member. The $2,500 award is given to one recipient annually.

GCA Awards For Summer Environmental Studies
Garden Club of America
598 Madison Avenue, 7th Floor
New York, NY 10022
Tel: 212-753-8287

For tuition, this $1,500 scholarship is awarded to twelve or more individuals annually.

LIFE Scholarship Program
Land Improvement Foundation for Education
1300 Maybrook Drive, PO Box 9
Maywood, IL 60153
Tel: 708-344-0700

To provide financial assistance for the study of natural resources and/or conservation in college. This program is open to high school seniors, high school graduates, and currently enrolled college students. Selection is based on academic record, career goals, and letters of recommendation. Applicants must be enrolled in or accepted at an accredited institution. The $1,000 award is given to three recipients annually.

Melville H. Cohee Student Leader Conservation Scholarship
Soil and Water Conservation Society
7515 Northeast Ankeny Road
Ankeny, IA 50021-9764
Tel: 515-289-2331 or 800-THE SOIL

To provide financial assistance to members of the Soil and Water Conservation Society (SWCS) who are interested in pursuing undergraduate or graduate studies with a natural resource conservation orientation. Applicants must have been members of the society for more than one year, have served for one academic year or longer as a student chapter officer for a chapter with at least fifteen members, have at least a 3.0 grade point average, be in the final year of undergraduate study or pursuing the first or second year of graduate study in conservation or related environmental protection or resource management fields at an accredited college or university, be in school at least half time, and not be an employee or immediate family member of the scholarship selection committee. Financial need is not considered in the selection process. The stipend is $900.

Morris K. Udall Scholarship Program
Web: http://www.act.org/udall

The foundation awards approximately seventy-five scholarships to outstanding students in their sophomore or junior year who have outstanding potential and intend to pursue careers in environmental public policy. Must be nominated by his or her college or university using the official nomination materials provided to each institution. Thus far, 125 undergraduate students have been awarded scholarships up to $5,000. Two $24,000 fellowships have been awarded to Ph.D. candidates.

National Council of State Garden Clubs, Inc. Awards
4401 Magnolia Avenue
St. Louis, MO 63110-3492

Available for students studying floriculture, horticulture, landscape design, city planning, land management, environmental concerns, and related subjects and who have endorsement of NCSGC in a related state. Only offered to undergraduate juniors, seniors, and graduate students. The $3,500 award is given to thirty-one recipients annually.

NEHA Scholarship Awards
National Environmental Health Association
720 South Colorado Boulevard, Suite 970
Denver, CO 80222
Tel: 303-756-9090

Must be accepted or enrolled in an environmental health program as an undergraduate. The award is between $400 and $1,000; the number of scholarships available varies.

NELA Scholarship Contest
Northeastern Loggers' Association, Inc.
PO Box 69
Old Forge, NY 13420-0069
Tel: 315 369-3078

To provide financial assistance to family members of the Northeastern Loggers' Association who are interested in studying forestry and wood science programs. This program is open to graduating high school seniors, students in two-year associate degree or technical school programs, and juniors and seniors in four-year baccalaureate programs. Applicants must be members of the immediate family of members of the association or immediate family of employees of industrial and associate members of the association. Selection is based primarily on a one-thousand-word essay on "What it means to grow up in the forest industry." Stipends are $500.

North American Loon Fund Grants
North American Loon Foundation
6 Lily Pond Road
Gilford, NH 03246
Tel: 603-528-4711

Applicants are chosen based upon approaches to studying capture techniques, migration, subadults, and winter ecology. Applicants should send a brief resume including education, experience, and any type of training acquired in the above fields. The award is between $500 and $3,000; the number of scholarships available varies, up to twenty.

Our World Underwater
PO Box 4428
Chicago, IL 60680
Tel: 312-666-6846

Applicants must be a certified scuba diver between the ages of eighteen and twenty-four, have an associate's or bachelor's degree with a good academic record, and pass a physical exam. Applicants should be interested in a career in marine science or related field. The $12,000 award is given to a single recipient.

Philip D. Reed Undergraduate Fellowship In Environmental Engineering
National Action Council for Minorities in Engineering
3 West 35th Street
New York, NY 10001-2281
Tel: 212-279-2626

To provide financial assistance for education in environmental engineering to underrepresented minority students. Engineering sophomores who are African-American, Hispanic, or Native American are eligible to be nominated by their deans for this award. Selection is based on high academic achievement and interest in environmental engineering. The stipend is $5,000 per year.

Plastics Recycling Competition
American Institute of Chemical Engineers
345 East 47th Street
New York, NY 10017-2395
Tel: 212-705-7478

To recognize and reward outstanding plastics recycling projects designed by undergraduate students. Undergraduate students in North America may enter this competition, whether or not they are members of the American Institute of Chemical Engineers (AICE). Acting as interdisciplinary teams, they design a large-scale, cost-effective plastics recycling system that converts used mixed plastics back into monomers or other marketable products. Entrants are required to consider technical, environmental, and economic issues in their solutions. First prize is $1,500, second $1,200, and third $900.

Seaspace Scholarship Program
Seaspace, Inc.
c/o Houston Underwater Club, Inc.
PO Box 3753
Houston, TX 77253-3753
Tel: 713-721-8533

To provide financial assistance to undergraduate and graduate students interested in preparing for a marine-related career. This program is open to junior, senior, and graduate students who are interested in preparing for a marine-related career. They should be majoring in marine science, marine biology, wildlife and fisheries, environmental toxicology, biological oceanography, genetics, ocean engineering, aquaculture, or zoology with marine mammal applications. Preference is given to graduate students. Selection is based on academic excellence (minimum grade point average of 3.5 for undergraduates or 3.0 for graduate students), demonstrated course direction, and financial need. The amount awarded varies each year. To date, financial assistance has been provided to more than 120 students.

Shakespeare Company Scholarships
National FFA Center
PO Box 15160
Alexandria, VA 22309
Tel: 703-360-3600

Must be a member of the Future Farmers of America. The award is between $500 and $10,000; the number of scholarships available varies.

Smithsonian Institution Environmental Research Center
PO Box 28
Edgewater, MD 21037
Tel: 410-798-4424

Academic and summer internships at the SIERC. Award includes stipend of $150 per week and free dorm space.

Youth Activity Fund
Explorers Club
46 East 70th Street
New York, NY 10021
Tel: 212-628-8383

Open to high school and undergraduate college students to help them participate in field research in the natural sciences anywhere in the world. Grants are to help with travel and expenses. The amount of each scholarship varies; the number of scholarships available varies.

Look to the Pros

FIRST

The following professional organizations offer a variety of materials, from career brochures to lists of accredited schools to salary surveys. Many of them also publish journals and newsletters that you should become familiar with. Many also have annual conferences that you might be able to attend. (While you may not be able to attend a conference as a participant, it may be possible to "cover" one for your school or even your local paper, especially if your school has a related club.)

When contacting professional organizations, keep in mind that they all exist primarily to serve their members, be it through continuing education, professional licensure, political lobbying, or just "keeping up with the profession." While many are strongly interested in promoting their profession and passing information about it to the general public, these busy professional organizations are not there solely to provide you with information. Whether you call your write, be courteous, brief, and to the point. Know what you need and ask for it. If the organization has a Web site, check it out first: what you're looking for may be available there for downloading, or you may find a list of prices or instructions, such as sending a self-addressed stamped envelope with your request. Finally, be aware that organizations, like people, move. To save time when writing, first confirm the address, preferably with a quick phone call to the organization itself, "Hello, I'm calling to confirm your address. . . ."

THE SOURCES

Air and Waste Management Association
PO Box 2861
Pittsburgh, PA 15230
Tel: 412-232-3444
Web: http://www.awma.org/education/studactv.htm

Check out the AWMA's Web site for their online Student Activities Manual, which lists a variety of environment-related activities and fundraising ideas. The Web site also lists colleges offering advanced degrees in the environmental sciences.

American Geological Institute
4220 Kings Street
Alexandria, VA 22303
Web: http://agi.umd.edu/agi/agi.html

AGI has been producing career-guidance information for many years, including the current brochure called *Careers in the Geosciences.* A new project is now in development, "Professional Career Pathways in the Geosciences," which will provide students at their career-decision stages (high school and college) with up-to-date information about career opportunities in the geosciences. There will also be an interactive electronic career service on Internet, which will include real-time dialogues with geoscientists and a posting of current employment opportunities.

American Society of Limnology and Oceanography
5400 Bosque Boulevard, Suite 680
Waco, Texas 76710-4446
800-929-2756
Fax: 817-776-3767

ASLO offers *Aquatic Science Career Information,* a free article available by mail or on-line at the Web site. Defines the field of aquatic science and discusses job opportunities, employment outlook, earnings, working conditions, and educational preparation. Also available is a list of college programs in aquatic science.

California Sea Grant College
University of California
9500 Gilman Drive, Department 0232
La Jolla, CA 92093-0232
Tel: 619-534-4444
Web: http://www-csgc.ucsd.edu/

CSGC offers *Marine Science Careers,* $5, 40 pages, a comprehensive guide to marine career areas that contains question and answer profiles, as well as photos of 38 marine scientists and other professionals in the field. Contact Sea Grant Communications, Kingman Farm, University of New Hampshire, Durham, NH 03824, for information on this publication. *Directory of Academic Marine Programs in California,* 82 pages, lists and describes for students, teachers, and counselors the marine programs at 48 two- and four-year schools of higher learning in California.

Ecological Society of America
Center for Environmental Studies
Arizona State University
Tempe, AZ 85287-3211
Tel: 602-965-3000
Web: http://www.sdsc.edu/~ESA/esa.htm

ESA offers *Careers in Ecology,* 8 pages.

Environmental Careers Organization
National Office
286 Congress Street, 3rd floor
Boston, MA 02210
Tel: 617-426-4375
Web: http://www.eco.org/

ECO provides career information, publications, and internship information, among many other services.

Environmental Protection Agency
Web: http://www.epa.gov/

According to the EPA, their mission "is to protect human health and to safeguard the natural environment—air, water, and land—upon which life depends." If it relates to the environment, they can help.

Marine Technology Society
1828 L Street, NW, Suite 906
Washington, DC 20036-5104
Tel: 202-775-5966
Web: http://www.electriciti.com/horrigan/mts/

MTS offers *University Curricula in Oceanography and Related Fields,* $6, 204 pages, which presents data (facilities, programs offered, faculty, student support, and contact information) on 266 colleges and universities, as well as 44 technical schools and institutions.

- **Mississippi-Alabama Sea Grant Consortium**
  PO Box 7000
  Ocean Springs, MS 39566-7000
  Tel: 601-875-9341
  Web: http://www.waidsoft.com/seagrant/

MASGC offers *Marine Education: A Bibliography of Educational Materials Available from the Nation's Sea Grant Programs,* $2.50, 51 pages, which lists Sea Grant programs and institutions and the materials they developed. Includes ordering instructions and information about materials available for free or at a nominal cost.

- **National Association of Conservation Districts**
  509 Capitol Court, NE
  Washington, DC 20002
  Tel: 202-547-6223
  Web: http://www.nacdnet.org

NACD offers *A Guide to Careers in Natural Resource Management,* 9 pages, which lists sources of career information in such fields as agriculture, biology, engineering, fisheries science, forestry, geology, landscape, marine science, range management, soil and water conservation, and wildlife management.

- **National Park Service**
  U.S. Department of the Interior
  Public Information Office
  PO Box 37127
  Washington, DC 20013-7127
  Tel: 202-208-5228
  Web: http://nps.gov/

The National Park Service offers *Careers in National Park Service,* 20 pages, which contains information about the National Park Service, describes career opportunities in the field, covers employment benefits and the application and hiring process, and lists Federal job information/testing offices.

National Science Foundation
4201 Wilson Boulevard
Arlington, Virginia 22230
Tel: 703-306-1234
Web: http://www.nsf.gov/

The NSF has a wide variety of programs for students of all ages. Contact them for education information, scholarships, fellowships, and their summer undergraduate research sites program.

The Nature Conservancy
1815 North Lynn Street
Arlington, VA 22209
Tel: 703-841-5300
Web: http://www.tnc.org

TNC offers *Careers at the Nature Conservancy,* 7 pages, which discusses career areas and how to apply for employment opportunities. *The Nature Conservancy Nation-Wide Internships/Short Term Positions,* 20 pages, lists and describes typical internships and short-term positions that recur annually.

North American Association for Environmental Education
Publications Department
6840 State Road 718
Pleasant Hill, OH 45359-9705
Tel: 202-884-8913
Web: http://eelink.umich.edu/naaee.html

NAAEE offers *List of Colleges and Universities with Programs Related to EE,* $6 plus $3.95 shipping, 35 pages, which lists by state the names and addresses of colleges and universities offering environmental education programs.

The Oceanography Society
4052 Timber Ridge Drive
Virginia Beach, Virginia 23455-7017
Tel: 757-464-0131
Web: http://www.tos.org/

Among other things the OS offers education and cross-disciplinary research opportunities to students (high school through college graduate), young investigators (post-graduate) and university faculty members. Also offers *Careers in Oceanography and Marine-Related Fields.*

Scripps Institution of Oceanography
8602 La Jolla Shores Drive
La Jolla, CA. 92037
Tel: 619-534-3624
Web: http://sio.ucsd.edu/

The Scripps Web site has a huge, must-see list of articles on careers in marine science by people who are in the field. There is also a list of volunteer and internship opportunities open to students.

Sea Grant College Program
Sea Grant Publications
1716 Briar Crest, Suite 603
Bryan, TX 77802
Tel: 409-862-3770
Web: http://texas-sea-grant.tamu.edu/

SGCP offers *Questions about Careers in Oceanography,* 18 pages, which is directed toward high school and college students, teachers, and guidance counselors with questions about careers in oceanography. Addresses what an oceanographer is, where to study oceanography, who hires oceanographers, who supports oceanographers, and where to obtain further information. *Vocational-Technical Marine Career Opportunities in Texas,* 24 pages, covers the maritime transportation industry (merchant marines, inlandmarine transportation, and offshore supply and transportation); the offshore mineral, oil, and gas industry; commercial diving; commercial fisheries; and shipbuilding.

Sea World of Florida
Education Department/Book Orders
7007 Sea World Drive
Orlando, FL 32821-8097
Tel: 407-363-2207
Web: http://www.seaworld.org

Sea World's Education Department offers more than 75 marine science educational resources (booklets, posters, and teacher's guides) designed for K-12 students and teachers. Write for a complete list and order form.

Society of American Foresters
5400 Grosvenor Lane
Bethesda, MD 20814-2198
Tel: 301-897-8720
Web: http://www.safnet.org

SAF offers *So You Want to Be in Forestry,* 16 pages (send self-addressed, stamped 9 X 12 envelope), which explains the roles and duties of foresters, their education and training, career opportunities, and related fields. The following publications are in the Forestry Career Packet: *Accredited Professional Forestry Degree Programs,* 2 pages, lists institutions with SAF-accredited curricula and SAF-recognized curricula in the United States and Canada; *Forestry Career Information Question and Answer Sheet,* 2 pages, contains the most frequently asked questions about the profession of forestry; *Job Seekers' Guide,* 2 pages, lists contact information of forestry employers.

Student Conservation Association, Inc.
1800 North Kent Street, Suite 1260
Arlington, VA 22209-2104
Tel: 703-524-2441
Web: http://www.sca-inc.org/

SCA offers *Conservation Career Development Program,* 5 information sheets, which describes the program that trains ethnic minorities and women at both the high school and college level for careers with natural and cultural resources management organizations. *Resource Assistance Program: Volunteer Positions,* 82 pages, lists volunteer opportunities by state for a wide range of natural and cultural resource assistance programs. *The Student Conservation Association: Making a Difference,* describes activities and programs you can participate in.

United States Fish and Wildlife Service
Office of Public Affairs
U.S. Department of the Interior
Washington, DC 20240
Tel: 202-208-5611
Web: http://www.fws.gov/

A Challenge and an Adventure. A video describing career opportunities with the Fish and Wildlife Service. To borrow a copy, contact your regional office or the Office of Public Affairs (202-208-5611) to find the branch nearest you. *Careers with the U.S. Fish and Wildlife Service,* 20 pages, has more color photos than text. This book highlights the problems addressed by the service (pollution, deforestation, wildlife habitat), and describes the academic background required for its jobs.

U.S. Geological Survey
807 National Center
Reston, VA 20192
Web: http://www.usgs.gov/

Contact the USGS or visit their Web site for information about their environment theme, including current and future projects, as well as a registry of Earth and Environmental Science Internet resources.

- **Wildlife Society**
  5410 Grosvenor Lane
  Bethesda, MD 20814-2197
  Tel: 301-897-9770
  Web: http://www.wildlife.org/index.html

The Wildlife Society offers *A Wildlife Conservation Career for You,* 12 pages, which describes careers in wildlife management, related opportunities, education needed, and personal requirements. *Universities and Colleges Offering Curricula in Wildlife Conservation,* 6 pages, lists North American campuses that have special curricula related to the fields of wildlife conservation and management.

- **Woods Hole Oceanographic Institution**
  Woods Hole, MA 02543
  Tel: 508-457-2000
  Web: http://www.whoi.edu/

Contact WHOI for everything you ever wanted to know about the ocean and oceanography.

Index

D

E

F

G